# Propagation of Horticultural Plants: Arid and Semi-Arid Regions

*Editor*

**Benja Rambeloson**

scitus academics

**Propagation of Horticultural Plants: Arid and Semi-Arid Regions**

Edited by **Benja Rambeloson**

ISBN: 978-1-68117-581-2
Library of Congress Control Number: 2016946436


SCITUS Academics LLC,
616, Corporate Way, Suite 2, 4766,
Valley Cottage, NY 10989

www.scitusacademics.com



Printed in United States of America on Acid Free Paper

# Preface

Over the past 40 years, the world's agricultural systems have been changing in response to population pressures. Population growth and local economics are driving both the intensification of agriculture and its extensification in to the marginal lands, where risks of crop failure and environmental degradation are high. Horticulture based production systems are now considered to be the most ideal strategy to provide food, nutrition and income security to the people. Integration of annual crops with fruit trees yields multiple outputs that ensure production and income generation. The importance of horticulture in improving the productivity of the land, generating employment, improving economic conditions of the farmers and entrepreneurs, enhancing exports and above all, providing nutritional security to the desert dwellers, can hardly be overemphasized. Horticulture has assumed significant importance in the crop diversification in recent years, which has become essential to arrest serious land degradation and enhancing the farm income. In fact, the horticulture has also gained commercial importance with a very significant share in the economy of the region. Diversification of agriculture from traditional land use with predominantly cereal/legume-based cropping systems to more productive and remunerative has become a milestone to be achieved. Horticulture provides the few viable and most attractive alternative land use system. The improved cultivars of prevalent fruit crops with their propagation method, spacing and yield potential was worked out during past four decades. Plant propagation is the branch of horticulture which deals with the deliberate production of new plants using various starter materials (e.g. organs, tissues), including their intensive but temporary care. It is primarily practiced to produce seedlings or clones of nursery crops for out-planting, or for planting in containers for display or decor or other uses.

*Propagation of Horticultural Plants: Arid and Semi-Arid Regions* deals with the propagation of fruit and plantation crops, of spice and medicinal plants. The book is aimed at post-graduate horticultural students, researchers and horticulturists.

# Table of Contents

# CHAPTER 1

# Response of Short Duration Tropical Legumes and Maize to Water Stress: A Glasshouse Study

**Hossain Sohrawardy[1] and Md. Lokman Hossain[2]**

[1]Department of Crop Production Systems in the Tropics,
Faculty of Agricultural Sciences, Georg-August-University,
37077 Göttingen, Germany

[2]Global Change Ecology (International Elite Graduate Programme),
Faculty of Biology, Chemistry and Geosciences, University of Bayreuth,
95440 Bayreuth, Germany

## ABSTRACT

The study was conducted as a pot experiment in the tropical glasshouse to evaluate the response of grain legumes (Phaseolus vulgaris, Vigna unguiculata, and Lablab purpureus) in comparison to maize (Zea mays) and estimate their potential and performance. Two experiments were established using completely randomized design. Physiological measurements (stomatal conductance, photosynthetic activities, and transpiration rates) were measured using LCpro instrument. Scholander bomb was used for the measurement of plant cell water potential. Significant difference was observed in different plant species with increase of different water regimes. Among the legumes, L. purpureus showed better response in water stressed conditions. At the beginning, in dry watered treatment the photosynthetic rate was below 0 μmol $m^{-2}$ $s^{-1}$ and in fully watered condition it was 48 μmol $m^{-2}$ $s^{-1}$. In dry treatment, total dry weight was 10 g/pot and in fully watered condition it was near to 20 g/pot in P. vulgaris. The study concludes that water stress condition should be taken into consideration for such type of crop cultivation in arid and semiarid regions.

## 1. INTRODUCTION

Water is the most common limiting factor to primary productivity in arid and semiarid areas [1]. Areas with higher precipitation in a region have more above-ground primary productivity [2, 3]. Water-use efficiency is often equated with drought resistance and the improvement of crop yield under stress conditions [4]. Water has an influence on grain yield and yield contributing parameters. Canopy development is sensitive to water deficit [5]. Water stress is one of the most important factors for legumes production. It affects not only the production of the grains but also the whole process of growth of all organs of the plants and metabolism [6]. Plant growth and development are affected by water stress affecting physiological and biochemical processes, for example, ion uptake, photosynthesis, respiration and translocation [7–11]. Environmental stresses such as drought, salinity, heat, and cold represent a significant constraint to meet the world food demand, for example, Islamic Republic of Iran, especially under low precipitation (often <250 mm per annum). Furthermore, the uneven temporal and spatial distributions have diode agronomists to pick the foremost effective irrigation strategies or drought tolerant cultivars [12].

Grain legumes are a major source of protein in arid and semiarid regions of the world and play a key role in the economies of arid and semiarid regions [13]. Multipurpose (dual purpose) legumes provide fodder, forage, green manure, seeds, and leaves that are used in human food [14]. Cowpea (Vigna unguiculata L.) and lablab (Lablab purpureus L.) widely cultivated in the tropics have multipurpose uses [15, 16]. L. purpureus has been comparatively neglected in analysis and development and consequently our information of its genetic diversity is proscribed [16, 17].

Legumes fix atmospheric nitrogen and are important to restore soil fertility. Legumes also play important role in breakdown of disease cycles, improve physical structure of soil, encourage microorganisms, and release available phosphorus to the soil [18].

Morphological and physiological understanding of legumes could be useful for selecting varieties to obtain better yield under drought conditions [19, 20]. The response of plants to water stress varies considerably and depends upon the intensity and duration of stress as well as of genotype and growth stages of the plant [10, 21]. Understanding plant response to drought is important and essential to develop drought tolerant crops [22, 23].

## 1.1. Grain Legumes

### 1.1.1. Common Bean (Phaseolus vulgaris L.)

P. vulgaris one of the most important legume species in the developing countries and three quarters of the world production (8.5 million tonnes) is cultivated annually [24]. A major constraint to bean production in many developing countries is drought, which affects 73% of the area planted to beans in Latin America [25]. More than 300 million people get protein, carbohydrates, and minerals from grain legumes such as V. unguiculata in these regions. P. vulgaris is the second important source of calories after Z. mays in many areas of the Sub-Saharan Africa [26]. P. vulgaris is primarily a food in sub-Saharan Africa. Furthermore, millions of small-scale farmers in Latin America and Africa rely on the production and sale of bean as an important source of household income.

### 1.1.2. Cowpea (Vigna unguiculata L.)

V. unguiculata is also one of the most important grain legumes in the diet of humans living in the tropics and subtropics. V. unguiculata is also a major crop in West and Central Africa for livelihood of millions of people. They use it as food and fodder and also for cash income. V. unguiculata grains contain protein which is not expensive and both rural people and urban consumers use it as a major protein source. V. unguiculata grain contains about 25% protein and 64% carbohydrate [27]. V. unguiculata is a legume crop which is grown in hot-dry tropics and subtropics. It plays a substantial role by serving as grain and vegetables crop for the rural people in the east, west, south, and central part of Africa [28]. It is truly a dual purpose grain legume crop, providing food for man and feed for livestock and serving as a valuable and dependable revenue-generating commodity for small holder farmers and grain traders [29, 30]. V. unguiculata is also a valuable component of farming systems where soil fertility is limiting because V. unguiculata has a high rate of nitrogen fixation [31] and forms effective symbiosis with mycorrhizae [32] and has the ability to better tolerate a wide range of soil pH compared to other grain legumes [33]. V. unguiculata is utilized in crop rotation and it improves soil fertility for yield of the cereal grains [34–36]. In the developing countries where soil infertility is high, rainfall is limiting, and most of V. unguiculata is grown without the use of fertilizers and plant protection measures (i.e., pesticides or herbicides), and a wide variety of biotic and abiotic constraints also limit growth and severely limit yield [37,38].

### 1.1.3. Lablab (Lablab purpureus L.)

L. purpureus is fast growing multipurpose legume which is grown in tropical region. Grain and forage types exist. It is suitable for ley and forage in Australia. L. purpureus originated from Africa but now it is cultivated in many tropical countries. It is grown as a grain legume crop in India and Africa. It is also adapted to grow in a wide range of rainfall and temperatures [39]. L. purpureus yields 4000 kg dry matter (DM)/ha of above-ground biomass in Australia under dry-land condition [40, 41]. Furthermore L. purpureus improves nitrogen (N) fertility of the soil [42,

43]. L. purpureus germination generally occurs within 5 days after planting. The growth period ranges between 75 and 300 days. Fruiting of some cultivars can be within 60–65 days of sowing and continues for 90–100 days. L. purpureus is adapted to drought condition [44] and known to be even more drought tolerant than P. vulgaris or V. unguiculata [45].

### 1.1.4. Maize (Zea mays L.)

Z. mays is one of the important cereal crop species after wheat and rice. It grows throughout a wide range of climates [46]. Z. mays shows susceptibility to drought in every stage but these three stages are more crucial for drought stress, that is, early growth stage, flowering stage, and mid-to-late grain filling stage [47]. Although Z. mays makes efficient use of water, it is considered more susceptible to water stress than other crops because of its unusual floral structure with separate male and female floral organs and the near-synchronous development of florets on a single ear borne on each. Low wet dry spore grains reduce the silk receptiveness and spore viability. Z. mays has completely different responses to water deficit consistent with development stages [48]. Drought stress is particularly damaging to grain yield if it occurs early in the growing season at flowering and during mid-to-late grain filling [49]. During the seedling stage, water stress damages secondary root growth and development. Stem growth and development after floral initiation show rapidly in adequate amount of water. Plants will be shorter and will decrease individual and cumulative area in water stress condition [50]. Ten to fourteen days before and after flowering of the Z. mays is the critical time for water stress. During the flowering stage grain yield reduces two to three times than any other stages of the growing period of Z. mays[51]. Z. mays grain yield is correlated with kernel number per plant during the water stress at flowering and grain filling stages [52], highlighting the importance of adequate water supplies during flowering.

The study was conducted to evaluate the drought tolerance of grain legumes compared to Z. mays by measuring agronomic (biomass production, leaf area, and yield) and physiological parameters (stomata conductance and photosynthetic activity) and to identify phenotypic variability of three commonly grown grain legume species in semi-arid Africa compared with the common staple cereal crop Z. mays in Africa.

# 2. MATERIALS AND METHODS

## 2.1. Materials

In experiments, common bean (P. vulgaris, KATX 56), cowpea (V. unguiculata, M 66), lablab (L. purpureus, Q 6880B), and maize (Z. mays, Ronaldinio KWS) were used in a glasshouse pot experiment (Figure 1). Figure 2shows the soil used for the pot experiment.

**Figure 1.** Seed of 4 species used in the experiment.

**Figure 2.** Soil used for the pot experiment.

## 2.2. Methods

Two glasshouse pot experiments were conducted as in the tropical glasshouse at Department of Crop Production Systems in the Tropics, Georg August University Göttingen, Göttingen, Germany. Experiment 1 was set using CRD with 3 different water regimes as treatments and replicated four times. Four species (P. vulgaris, V. unguiculata, L. purpureus, and Z. mays) were used in this experiment. Treatments were applied as (1) fully watered (watering three times in a week), (2) moderately watered (watering two times in a week), and (3) partially watered (watering one time in a week). Experiment 2 was set using CRD with 5 different water regimes as treatments and replicated three times. Three species (P. vulgaris, V. unguiculata, and L. purpureus) were used in this experiment. At first every pot was watered depending on weather condition. Treatments were applied as (1) fully watered (watering three times in a week), (2) moderately watered (watering two times in a week), (3) partially watered (watering one time in a week), (4) nearly dry (watering one time in every two weeks), and (5) dry (watering one time in every three weeks). Pot with a diameter of 23.5 cm and a depth of 19 cm with an internal capacity of approximately 8 lt. was used for both experiments. Pots were filled with substrate, clay, compost, and Schoninger sand (Figure 3). Soil pH (6.5) and bulk density (0.7 gm/cm$^3$) were measured before pot filling. Seeds were soaked in distilled water for 24 hours and planted into pots directly. Three seeds were planted in each pot at a depth of 4.0 cm approximately. The pots were watered to field capacity by repeated watering. After germination pots were thinned to one plant per pot, pots were subsequently watered according to the above described treatments from 3 weeks after planting. Soil volumetric water content was measured by soil moisture meter (TDR) in a week once before watering and after watering. Data were taken from 12 cm deep of the soil. Physiological measurements (stomatal conductance, photosynthetic activities, and transpiration) were taken using LCpro-SD, ADC, BioScientific Ltd. Data were analysed using Statistica software version 10 and MS excel 2007. Treatment effects and interactions were considered statistically significant at ≤0.05 level of probability.

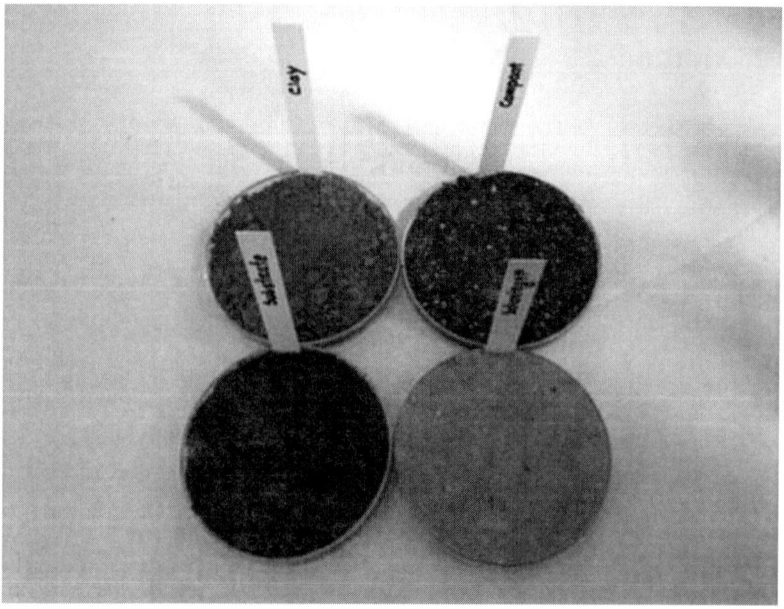


**Figure 3.** Components used for soil mix.

# 3. RESULTS

## 3.1. Cell Water Potential in Experiment 1

The water potential was significantly different for the different plant species tested in the experiment during the vegetative stage. It can be assumed that higher water potential of the cells had higher water stress. For the partially watered treatment water potential was the highest for all tested plant species. But the intensity of the responsiveness differed among plant species. Z. mays seemed to be stressed the most (high plant cell water potential for the partially watered treatment) followed by P. vulgaris, whereas the cell water potential remained stable for the legumes V. unguiculata and L. purpureus. Plant cell water potential of L. purpureus and V. unguiculata was 5 and 6 bar in partially watered condition. It is a good sign for drought tolerance. On the other hand, water potential of P. vulgaris and Z. mays was 11 and 17 bar, respectively, in fully watered condition (Figure 4).

## 3.2. Dry Weight of P. vulgaris, V. unguiculata, L. purpureus, and Z. mays in Experiment 1

Higher water regimes showed higher dry weight in all legumes and they also showed significant difference between different water levels. But interaction between different plants and water regimes did not show significant difference. Total dry weight was about 43 g/pot in fully watered condition in common bean, whereasV. unguiculata and L. purpureus had 40 g/pot dry weight. Besides, partially watered condition V. unguiculatashowed the highest value than other legumes, but L. purpureus showed the highest value in moderately watered treatment. This means that V. unguiculata and L. purpureus had low water use efficiency than P. vulgaris and partially watered condition V. unguiculata showed better yield than other legumes. On the other hand, L. purpureus had about 35 g/pot dry weight in partially watered treatment. Water stress affects total dry in P. vulgaris. In partially watered treatment, total dry weight was 20 g/pot in P. vulgaris. P. vulgaris showed above 40 g/pot total dry weight fully watered condition (Figure 5).

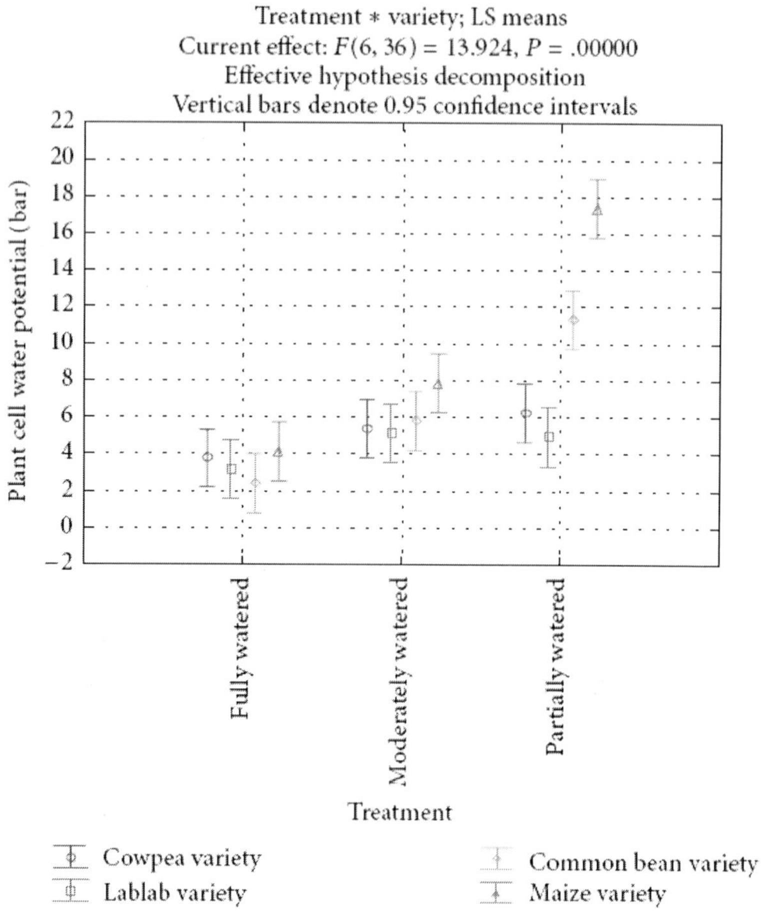


**Figure 4.** Plant cell water potential of P. vulgaris, V. unguiculata, L. purpureus, and Z. maysunder different treatments in experiment 1.

## 3.3. Stem-Leaf Ratio of V. unguiculata, L. purpureus, P. vulgaris, and Z. mays in Experiment 1

Different water regimes showed significant difference and different species also showed significant difference but interaction between variety and water regimes did not show significant difference. L. purpureus showed lowest stem and leaf ratio in fully watered condition in comparison to the moderately and partially watered treatments. This means that L. purpureus has low water use efficiency. On the other hand, non-legume crop, Z. mays, has more water use efficiency and water level was increased and hence, stem and leaf ratio showed higher value. In fully watered conditions, Z. mays had stem and leaf ratio was about 4.6 (Figure 6) than other watering treatments.

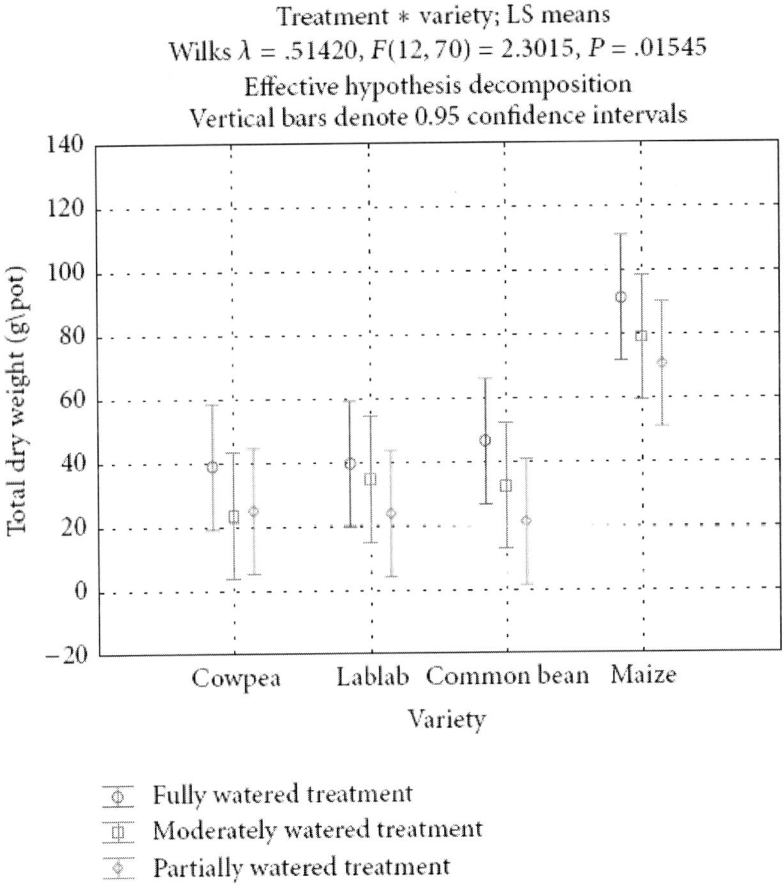


**Figure 5.** Total dry weight in different plants according to different watering treatments.

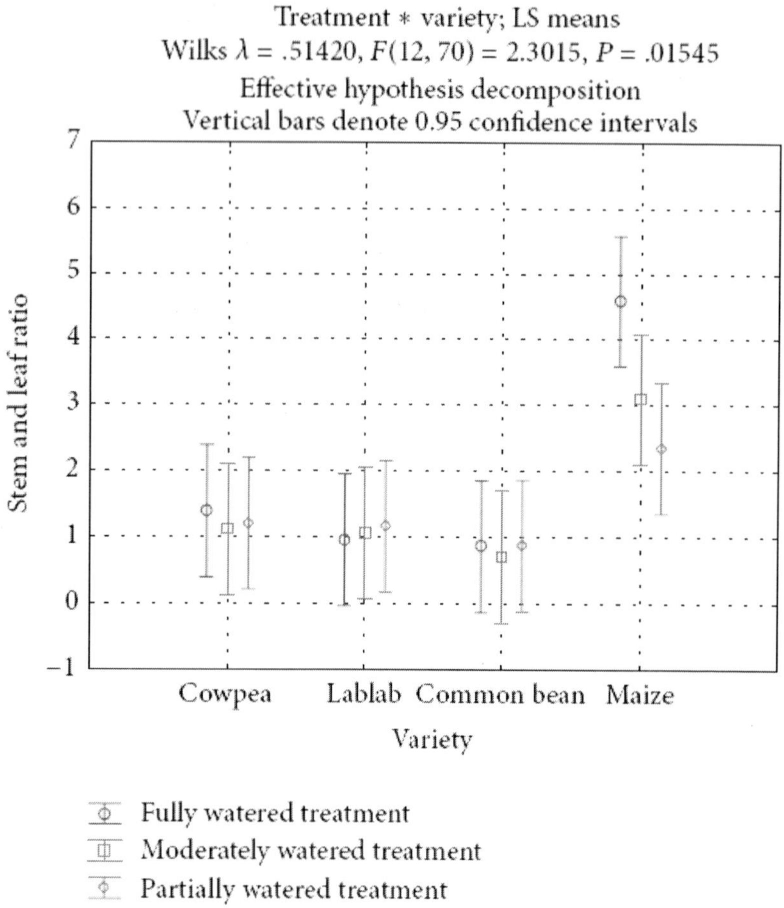


**Figure 6.** Stem-leaf ratio in different legumes according to different watering treatments.

## 3.4. Soil Moisture Content in Different Treatments of Four Species

Differences between crops in the same watering regime were often small except fully watered condition. In case of dry, nearly dry, and moderately watered conditions, water consumption of L. purpureus, V. unguiculata, andP. vulgaris wasmuch more similar, but in fully watered conditions, P.

vulgaris uptake more water than other crops. V. unguiculata and L. purpureus consumed less water in fully watered condition (Figure 7).

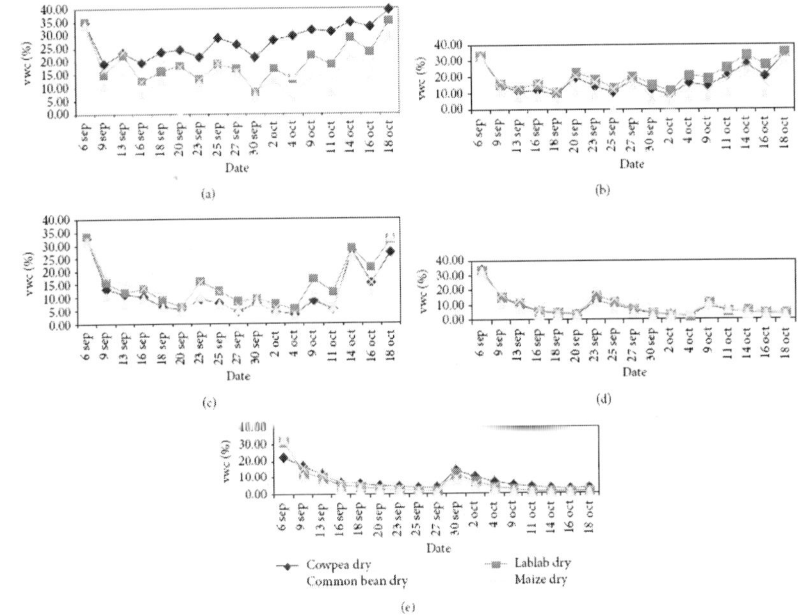


**Figure 7.** Soil moisture content of different water level: (a) fully watered (b) partially watered (c) moderately watered, (d) near to dry, and (e) dry.

## 3.5. Photosynthesis of V. unguiculata, L. purpureus, and P. vulgaris in Experiment 2

Photosynthesis showed significant difference in variety and treatment interaction between before and after watering. Before watering, V. unguiculata showed the highest photosynthesis rate compared to the other plant species tested in dry condition. L. purpureus showed gradually increase of photosynthesis before and after watering (Figures 8 and 9). Photosynthesis was little bit increased in P. vulgaris and decreased in L. purpureusand V. unguiculata through watering. This also indicated that L. purpureus had lower water use efficiency thanP. vulgaris.

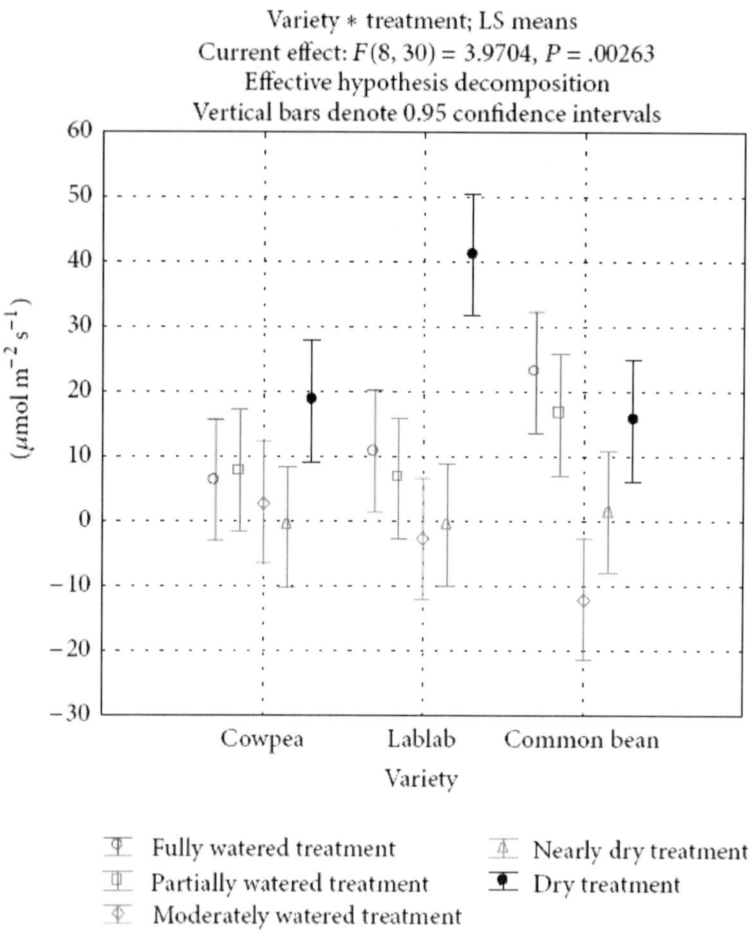


**Figure 8.** Photosynthesis rate on 55 DAP.

## 3.6. Stomatal Conductance of V. unguiculata, L. purpureus, and P. vulgaris in Experiment 2

Stomatal conductance showed significant difference in different water regimes in legumes. In P. vulgaris, stomatal conductance was increased when the water level increased. P. vulgaris showed more water use efficiency and stomatal conductance was gradually increased with the increase of water regimes. L. purpureusand V. unguiculata did not show too much variation; only L. purpureus in dry treatment suddenly showed very high stomatal conductance. L. purpureus and V. unguiculata showed less water use efficiency than P. vulgaris. After watering, it showed that P. vulgaris showed higher stomatal conductance in nearly dry, moderately watered, and partially watered treatments. In L. purpureus, stomatal conductance was increased in case of nearly dry and moderately watered conditions but decreased little bit in fully watered and partially watered conditions. After watering, V. unguiculata showed water use efficiency in nearly dry and moderately watered conditions, but there was no change in fully watered and partially watered conditions (Figures 10 and 11).

## 3.7. Transpiration of V. unguiculata, L. purpureus, and P. vulgaris in Experiment 2

Water regimes showed significant difference in transpiration before and after water use in different plants. In P. vulgaris, transpiration increased with the increase of water level except in dry treatment. After watering, evaporation increased slightly with the increase of water level in nearly dry and moderately watered conditions in L. purpureus, but transpiration was increased in partially and fully watered condition. In V. unguiculata, transpiration was increased in nearly dry and moderately watered conditions but decreased in partially and fully watered conditions on 55 DAP (Figures 12 and 13).

## 3.8. Photosynthesis of V. unguiculata, L. purpureus, and P. vulgaris on 55 DAP and 58 DAP in Experiment 2

Photosynthesis rate showed significant differences between different varieties and watering treatments on 55 DAP and 58 DAP. On 58 DAP, all varieties showed the highest photosynthesis rate in fully watered condition. On 55 DAP, P. vulgaris showed that increase of water level gradually increased photosynthetic rate. At the beginning, in dry watered treatment the photosynthetic rate was below 0 µmol $m^{-2}$ $s^{-1}$ and finally in fully watered condition it was 48 µmol $m^{-2}$ $s^{-1}$. Water level was increased followed by the increase of photosynthetic rate in fully watered condition. V. unguiculata and L. purpureus showed response to water level very well, but water use efficiency was lower in case of V. unguiculata and L. purpureusthan compared to P. vulgaris (Figures14 and 15).

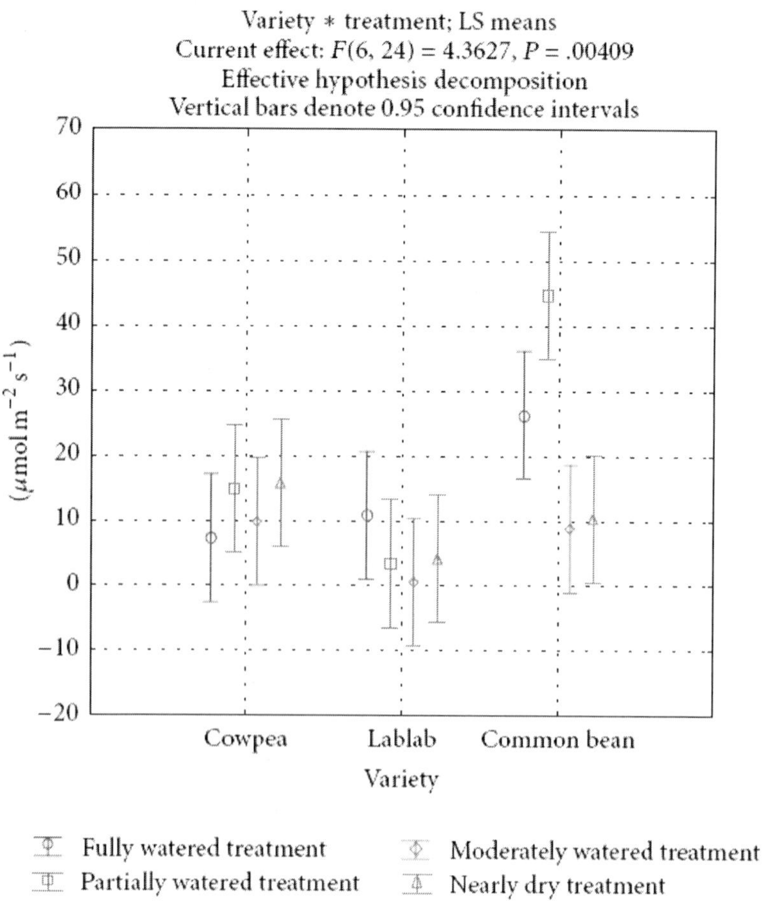


**Figure 9.** Photosynthesis rate on 58 DAP.

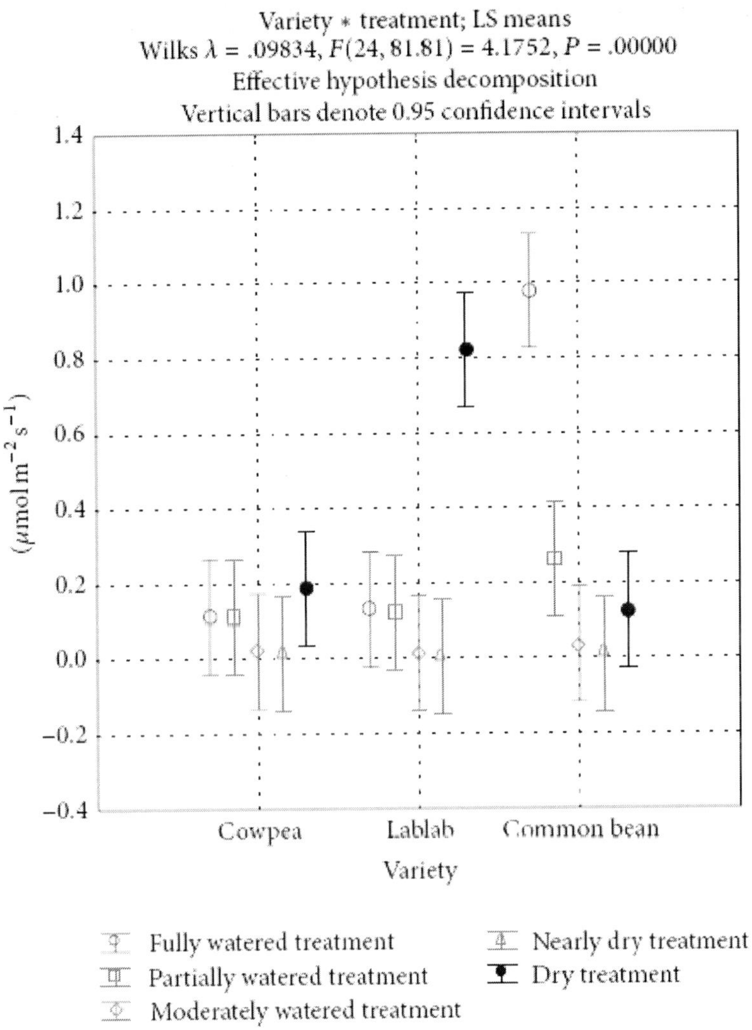


**Figure 10.** Stomatal conductance on 55 DAP.

## 3.9. Stomatal Conductance of V. unguiculata, L. purpureus, and P. vulgaris on 55 DAP and 58 DAP in Experiment 2

On 55 DAP, stomatal conductance showed significant difference in different legumes at different water regimes, but on 58 DAP stomatal conductance did not show significant difference. Stomatal conductance was gradually increased in V. unguiculata except for dry water treatments on 55 DAP (Figures 16 and 17).

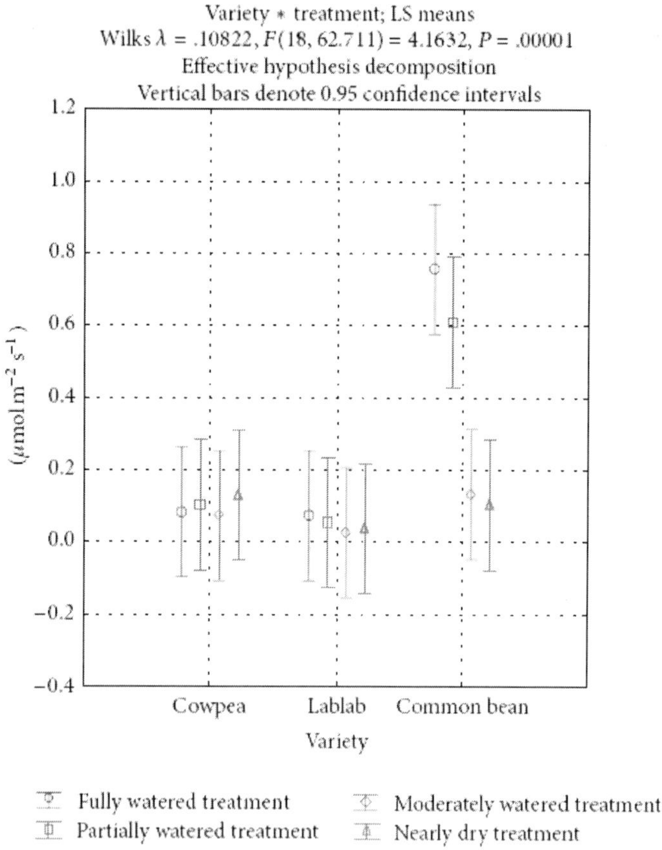


**Figure 11.** Stomatal conductance on 58 DAP.

## 3.10. Transpiration of V. unguiculata, L. purpureus, and P. vulgaris on 55 DAP and 58 DAP in Experiment 2

Transpiration showed significant differences with different water treatments on 55 DAP and 58 DAP in different plants. P. vulgaris showed the gradual increases of transpiration with increasing water levels on 55 DAP. L. purpureus and V. unguiculata showed higher transpiration in fully watered condition on 55 DAP, but other treatments showed very little variation. P. vulgaris had more water use efficiency on 55 DAP. P. vulgarisalso showed gradual increase of evaporation rate in different water regimes on 58 DAP. On 58 DAP, transpiration rate was increased after watering in P. vulgaris (Figures 18 and 19).

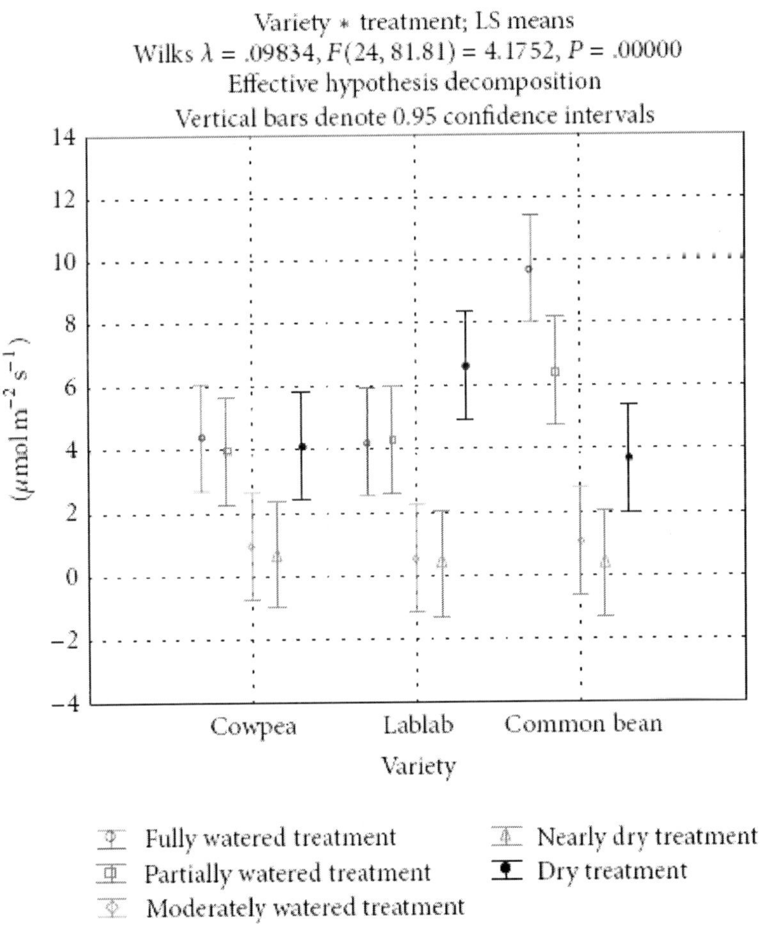


**Figure 12.** Transpiration on 55 DAP.

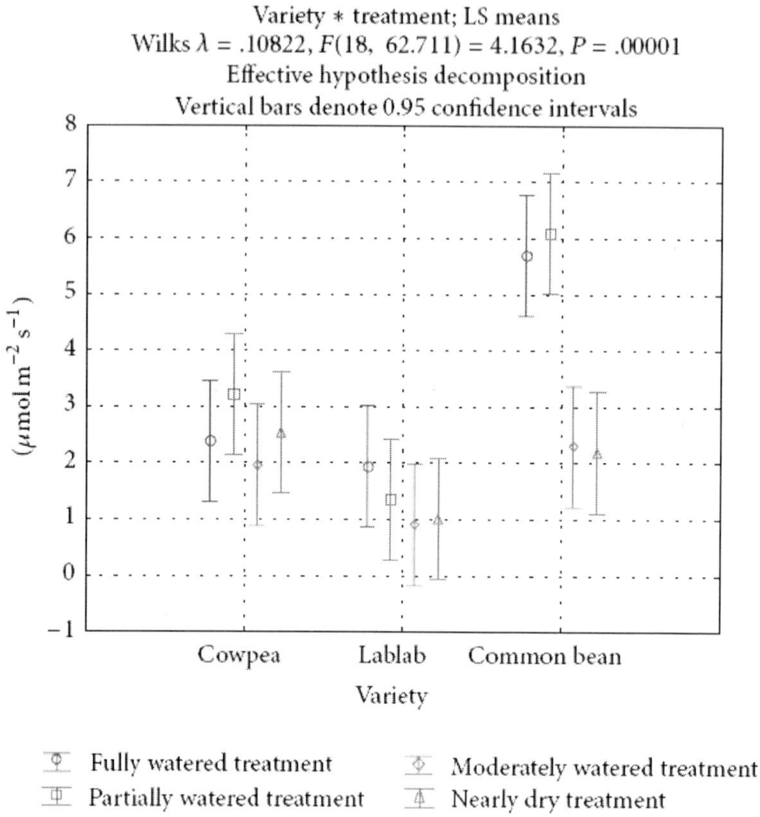


**Figure 13.** Transpiration on 58 DAP.

## 3.11. Stem-Leaf Ratio of V. unguiculata, L. purpureus, and P. vulgaris in Experiment 2

Different water regimes and species followed significant difference and also the interaction between species and different water level showed significant difference. P. vulgaris showed the lowest stem leaf ratio in comparison to the other legume species in different water regimes. V. unguiculata had the highest stem leaf ratio compared to the other crops in fully watered condition. There was a clear effect of the watering treatment on the leaf stem ratio. The leaf stem ratio was increased for all plant species with increased water availability. Different plant species had different water use efficiency. But it was also clear that availability of water was responsible for increased stem and leaf ratio (Figure 20).

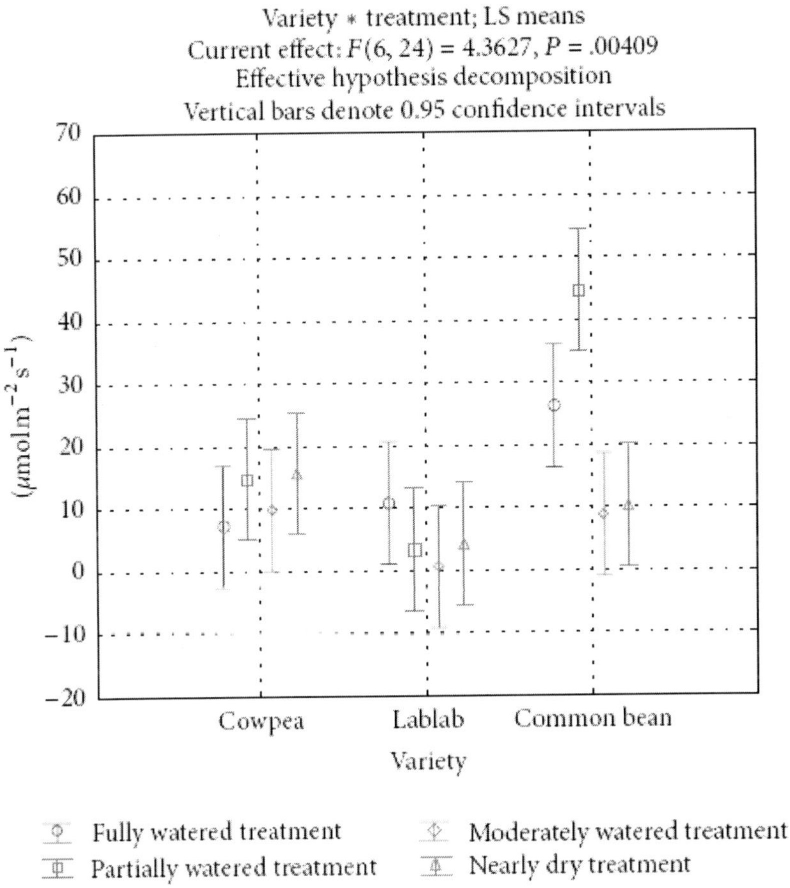


**Figure 14.** Photosynthesis on 55 DAP.

## 3.12. Total Dry Weight of V. unguiculata, L. purpureus, and P. vulgaris in Experiment 2

The effect of different water treatments on plant biomass production and dry weight was different among the three different legume species. For

P. vulgaris, dry weight was decreased with the increased water stress. In dry treatment, total dry weight was 10 g/pot and in fully watered condition it was near to 20 g/pot in P. vulgaris. Total dry weight was increased double in P. vulgaris when water level was higher. For L. purpureus the effect of different water treatments on dry weight was not pronounced and remained almost constant. Surprisingly, V. unguiculata showed the highest dry weight in the moderately watered plants. Moderately watered condition showed the highest dry weight compared to the other treatments (Figure 21).

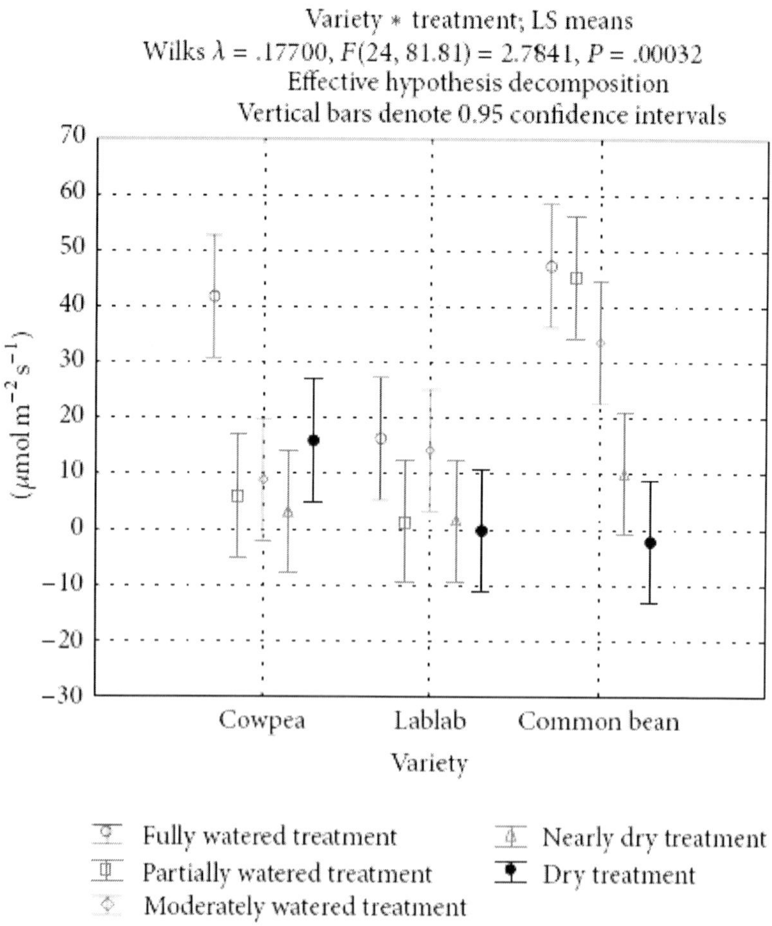


**Figure 15.** Photosynthesis rate on 58 DAP.

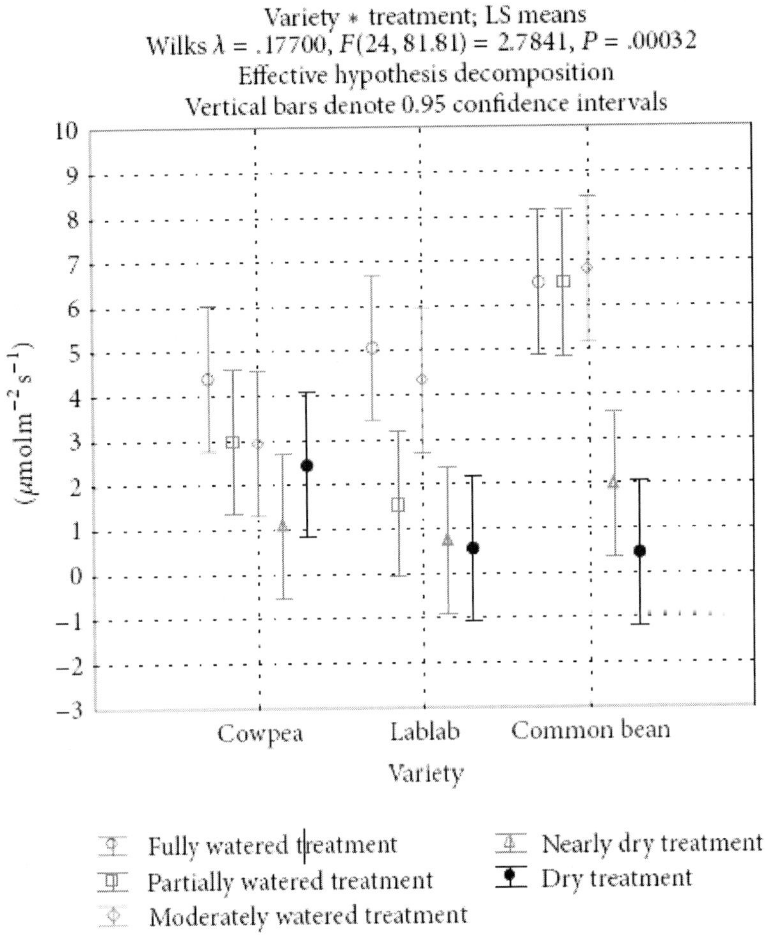


**Figure 16.** Stomatal conductance on 55 DAP.

# 4. DISCUSSIONS

L. purpureus showed gradually increased photosynthetic rate with response to different water level. In P. vulgaris, stomatal conductance was increased when the water level was increased and in fully watered condition it showed almost 1 $\mu mol\, m^{-2} s^{-1}$. L. purpureus and V. unguiculata did not show too much variation. Stomatal conductance was highest in P. vulgaris. L. purpureus, V. unguiculata, and P. vulgaris showed 0.0 μmol $m^{-2}$ $s^{-1}$stomatal conductance in nearly dry watering treatment, but in fully watered treatment stomatal conductance uplifted to 0.1 μmol $m^{-2}$ $s^{-1}$ in V. unguiculata and L. purpureus each, and P. vulgaris gave value near to 1.0 μmol $m^{-2}$ $s^{-1}$. The finding is also in agreement with Vogel et al. [53]. Among the legumes, L. purpureus showed lower photosynthetic rate than the other legumes. P. vulgaris had more water use efficiency than other legumes and stomatal conductance was gradually increased in regard to higher water regimes. L. purpureus and V. unguiculata did not show too much variation. L. purpureus and V. unguiculata followed less water use efficiency. In L. purpureus, stomatal conductance was increased in case of nearly dry and moderately watered conditions but decreased slightly in fully and partially watered conditions. The stomatal conductance was above 0.7 μmol $m^{-2}$ $s^{-1}$ in fully watered treatment and 0.1 μmol $m^{-2}$ $s^{-1}$ in partially watered treatment. On the other hand, V. unguiculata and P. vulgaris did not show too much change in fully watered treatment. Photosynthetic rate in P. vulgaris was increased with the increase of water level on 55 DAP. In dry watered treatment the photosynthetic rate was below 0 μmol $m^{-2}$ $s^{-1}$, but in fully watered condition it was 48 μmol $m^{-2}$ $s^{-1}$. Moreover, V. unguiculataand L. purpureus also showed positive response to water level but lower water use efficiency than P. vulgaris. P. vulgaris showed the gradual increases of transpiration with increasing water levels on 55 DAP. L. purpureus andV. unguiculata showed higher transpiration in fully watered conditions on 55 DAP. After watering the plants, P. vulgaris also showed gradually increases of transpiration rate in different water regimes on 58 DAP. It is proved that after watering transpiration rate was raised in all legumes, P. vulgaris showed clear gradual increase of transpiration than the other legumes. Higher water regimes showed higher dry weight in all legumes and they also showed significant difference between different water levels. Total dry weight in common bean was more (42 g/pot) in fully watered condition. V. unguiculata and L. purpureus had 40 g/pot dry weight. Besides, in partially watered condition, V. unguiculata showed the highest value compared to the other legumes. On the other hand, L. purpureus had about 35 g/pot dry weight. In partially watered treatment, total dry weight was 20 g/pot in P. vulgaris. Yield was reduced in P. vulgaris due to drought treatments. The water potential was significantly different for the different plant species tested in the experiment during the vegetative stage. It can be assumed that higher water potential of the cells belongs to higher water stress. For the partially water treatment water potential was highest for all tested plant species. But the intensity of the responsiveness differed among plant species. Z. mays needed adequate amount of water for its growth and development. Different legumes also showed different responses to water stress. Z. mays seemed to be stressed the most (high plant cell water potential for the partially water treatment) followed by beans, whereas the cell water potential remained stable for the legumes V. unguiculata and L. purpureus. Plant cell water potential of L. purpureus and V. unguiculata was 5 and 6 bar in partially watered condition. It is a good sign for drought tolerance.

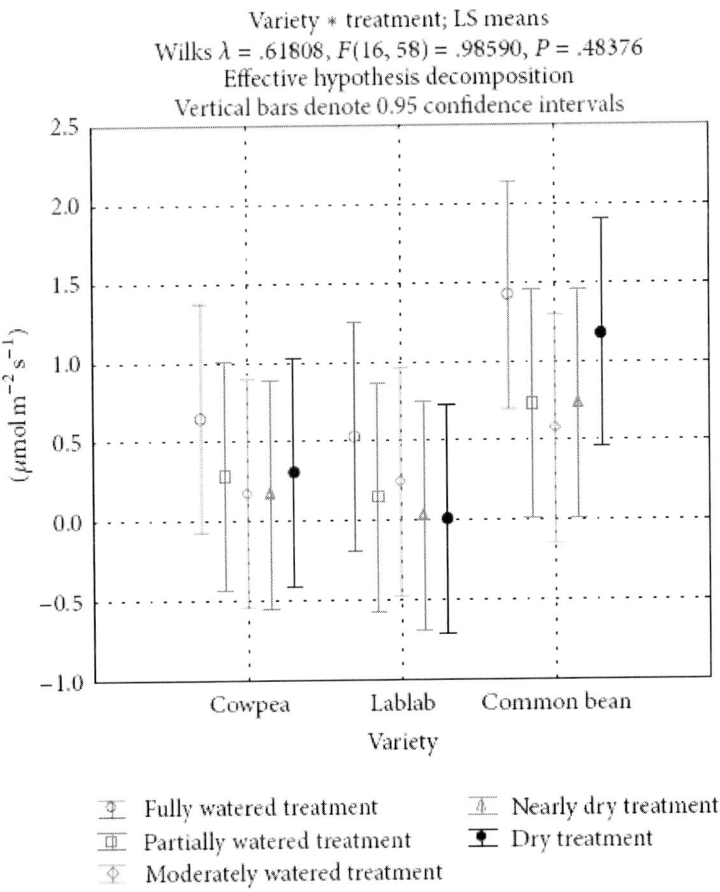


**Figure 17.** Stomatal conductance on 58 DAP.

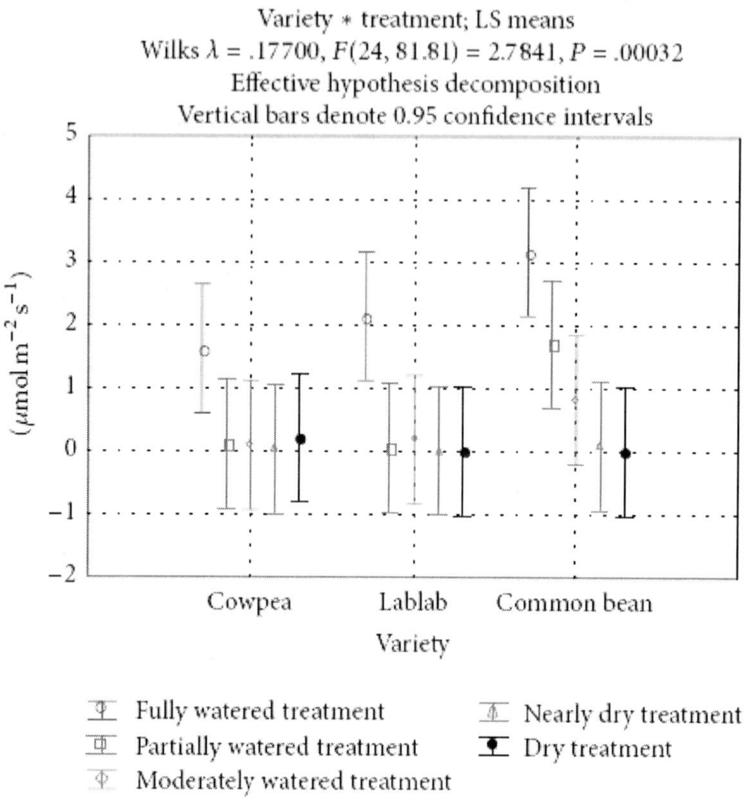


**Figure 18.** Transpiration on 55 DAP.

## 5. CONCLUSION

Water is the most limiting factor for agricultural production in particular in the semiarid areas of the world. Legumes are of great importance for food security particularly in the developing world. Furthermore, they have the ability to improve soil fertility in resource constraint small holder farming systems. Drought is the most limiting environmental factor for the productivity of crops in primarily rain-fed based farming systems of semiarid areas. Plant developed totally different methods to avoid or minimize drought stress. The study examines responsiveness of three legumes and Z. mays to water stress. Significant difference was observed in different plant species with increase of different water regimes. L. purpureus showed better response in water stress compared to the other species. P. vulgaris showed a negative effect on dry weight with increased water stress. Therefore, this study concludes that selection of crops in water stress condition is highly necessary to improve the crop production and ensure the agricultural and environmental sustainability.

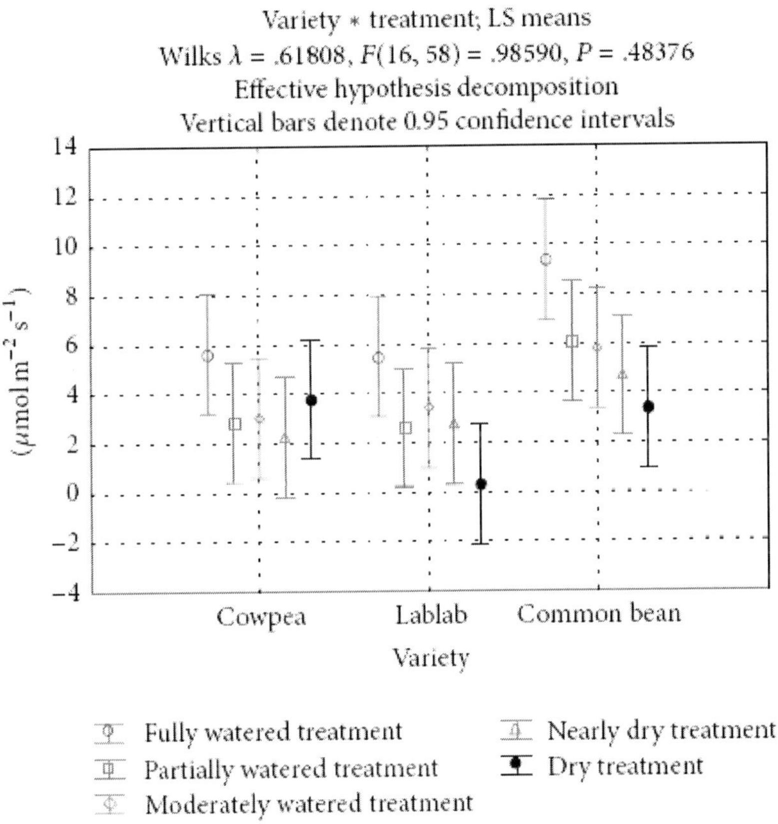


**Figure 19.** Transpiration on 58 DAP.

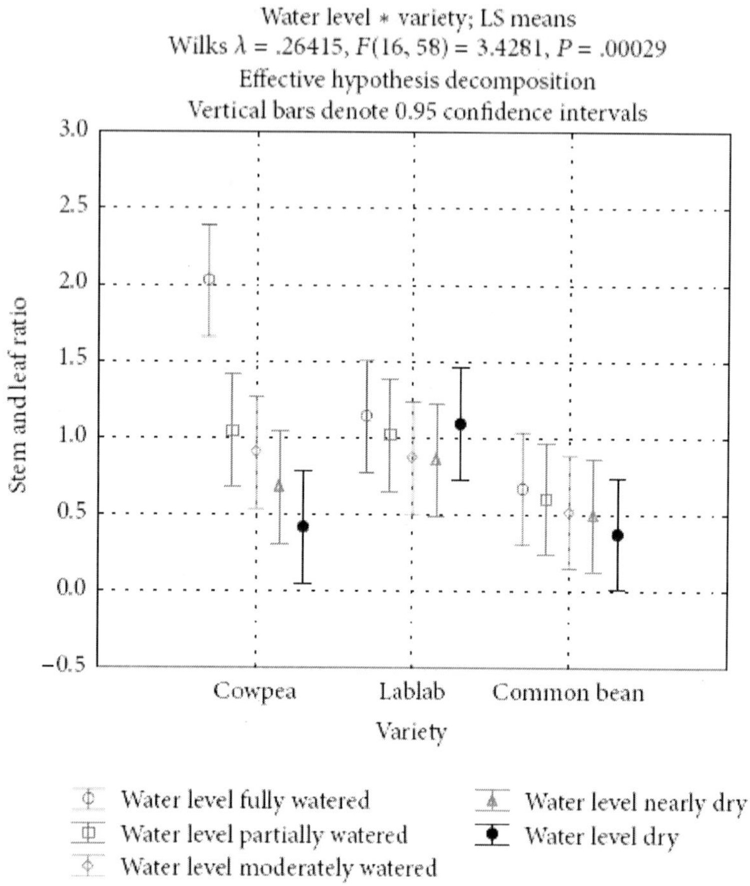


**Figure 20.** Stem-leaf ratio in different plants according to different watering treatments.

## ACKNOWLEDGMENTS

The authors extend their profound thanks to Professor Dr. Anthony Whitbread, PD Martin Worbes, Anne Sennhenn (Ph.D. student), and colleagues of the Department of Crop Sciences, Crop Production Systems in the Tropics, Georg August University Göttingen, Germany, for their assistance during the research work.

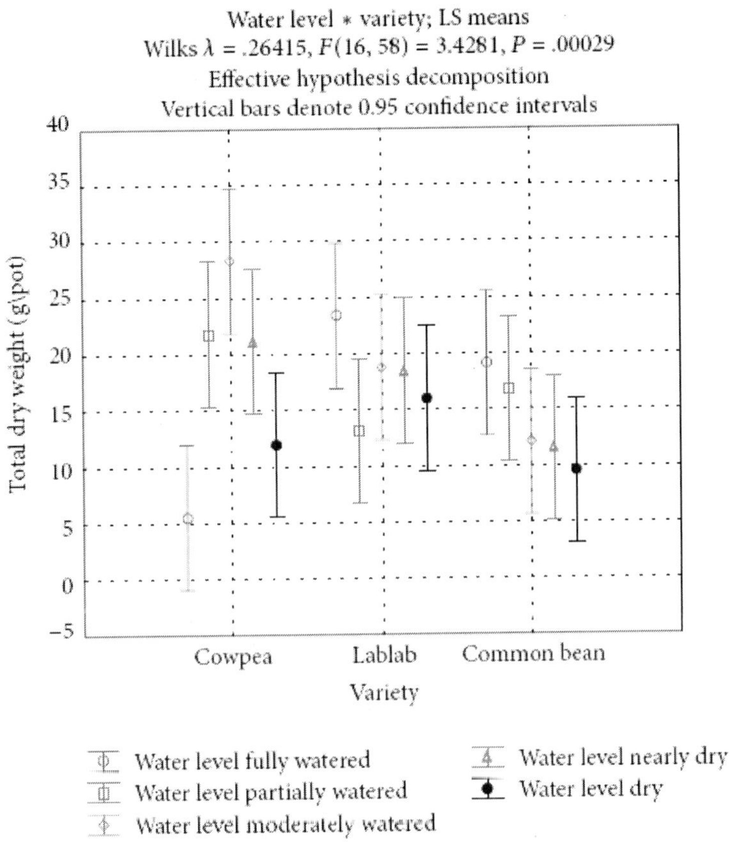


**Figure 21.** Total dry weight in different plants according to different watering treatments.

## REFERENCES


1. J. H. Brown, T. J. Valone, and C. G. Curtin, "Reorganization of an arid ecosystem in response to recent climate change," Proceedings of the National Academy of Sciences of the United States of America, vol. 94, no. 18, pp. 9729–9733, 1997.

2. M. L. Rosenzweig, "Net primary productivity of terrestrial communities: prediction from climatological data," The American Naturalist, vol. 102, no. 923, pp. 67–74, 1968.

3. O. E. Sala, W. J. Parton, L. A. Joyce, and W. K. Lauenroth, "Primary production of the central grassland region of the United States," Ecology, vol. 69, no. 1, pp. 40–45, 1988.

4. A. Blum, "Drought resistance, water-use efficiency, and yield potential—are they compatible, dissonant, or mutually exclusive?" Australian Journal of Agricultural Research, vol. 56, no. 11, pp. 1159–1168, 2005.

5. T. C. Hsiao, "The soil-plant-atmosphere continuum in relation to drought and crop production," inDrought Resistance in Crops with Emphasis on Rice, pp. 39–52, IRRI, Los Banos, Philippines, 1982.

6. M. J. Zimmermann, M. Rocha, and T. Yolanda, Bean Crop: Factors Affecting Productivity for the Presence of Potash and Phosphate, 1988.

7. C. A. Jaleel, R. Gopi, B. Sankar, M. Gomathinayagam, and R. Panneerselvam, "Differential responses in water use efficiency in two varieties of Catharanthus roseus under drought stress," Comptes Rendus Biologies, vol. 331, no. 1, pp. 42–47, 2008.

8. C. A. Jaleel, P. Manivannan, G. M. A. Lakshmanan, M. Gomathinayagam, and R. Panneerselvam, "Alterations in morphological parameters and photosynthetic pigment responses of Catharanthus roseusunder soil water deficits," Colloids and Surfaces B: Biointerfaces, vol. 61, no. 2, pp. 298–303, 2008.

9. C. A. Jaleel, R. Gopi, P. Manivannan, M. Gomathinayagam, R. Sridharan, and R. Panneerselvam, "Antioxidant potential and indole alkaloid profile variations with water deficits along different parts of two varieties of Catharanthus roseus," Colloids and Surfaces B: Biointerfaces, vol. 62, no. 2, pp. 312–318, 2008.

10. C. A. Jaleel, R. Gopi, and R. Panneerselvam, "Growth and photosynthetic pigments responses of two varieties of Catharanthus roseus to triadimefon treatment," Comptes Rendus—Biologies, vol. 331, no. 4, pp. 272–277, 2008.

11. M. Farooq, S. M. A. Basra, A. Wahid, Z. A. Cheema, M. A. Cheema, and A. Khaliq, "Physiological role of exogenously applied glycinebetaine to improve drought tolerance in fine grain aromatic rice (Oryza sativa L.)," Journal of Agronomy and Crop Science, vol. 194, no. 5, pp. 325–333, 2008.

12. A. Soltani and A. Faraji, Soil Water and Plant Relationship, Mashhad University, Mashhad, Iran, 2007.

13. J. Singh and A. Patal, "Water statues, gaseous exchange, proline accumulation and yield of mung bean in response to water stress," Annual of Biology Ludhiana, vol. 12, pp. 77–81, 1996.

14. A. I. Moalafi, J. A. N. Asiwe, and S. M. Funnah, "Germplasm evaluation and enhancement for the development of cowpea (Vigna unguiculata L.) dual-purpose $F_2$ genotypes," African Journal of Agricultural Research, vol. 5, no. 7, pp. 573–579, 2010.

15. B. B. Singh, H. A. Ajeigbe, S. A. Tarawali, S. Fernandez-Rivera, and M. Abubakar, "Improving the production and utilization of cowpea as food and fodder," Field Crops Research, vol. 84, no. 1-2, pp. 169–177, 2003.

16. B. C. Pengelly and B. L. Maass, "Lablab purpureus (L.) Sweet—diversity, potential use and determination of a core collection of this multi-purpose tropical legume," Genetic Resources and Crop Evolution, vol. 48, no. 3, pp. 261–272, 2001.

17. B. L. Maass, R. H. Jamnadass, J. Hanson, and B. C. Pengelly, "Determining sources of diversity in cultivated and wild Lablab purpureus related to provenance of germplasm by using amplified fragment length polymorphism," Genetic Resources and Crop Evolution, vol. 52, no. 6, pp. 683–695, 2005.

18. K. Hoshikawa, "Significance of legume crops in improving the productivity and stability of cropping systems," in Phosphorus Nutrition of Grain Legumes in the Semi Arid Tropices, C. Johannesen, K. K. Lee, and K. L. Sahrawat, Eds., pp. 173–181, International Crops Research Institute for the Semi-Arid Tropics, Paatancheru, India, 1991.

19. N. H. Nam, Y. S. Chauhan, and C. Johansen, "Effect of timing of drought stress on growth and grain yield of extra-short-duration pigeonpea lines," Journal of Agricultural Science, vol. 136, no. 2, pp. 179–189, 2001.

20. J. P. Martínez, H. Silva, J. F. Ledent, and M. Pinto, "Effect of drought stress on the osmotic adjustment, cell wall elasticity and cell volume of six cultivars of common beans (Phaseolus vulgaris L.)," European Journal of Agronomy, vol. 26, no. 1, pp. 30–38, 2007.

21. M. M. Chaves, J. S. Pereira, J. Maroco, et al., "How plants cope with water stress in the field. Photosynthesis and growth," Annals of Botany, vol. 89, pp. 907–916, 2002.

22. A. R. Reddy, K. V. Chaitanya, and M. Vivekanandan, "Drought-induced responses of photosynthesis and antioxidant metabolism in higher plants," Journal of Plant Physiology, vol. 161, no. 11, pp. 1189–1202, 2004.

23. C.-X. Zhao, L.-Y. Guo, C. A. Jaleel, H.-B. Shao, and H.-B. Yang, "Prospectives for applying molecular and genetic methodology to improve wheat cultivars in drought environments," Comptes Rendus: Biologies, vol. 331, no. 8, pp. 579–586, 2008.

24. D. Pachico, "Trends in world P. vulgaris production," H. F. Schwartz and M. A. Pastor Corrales, Eds., pp. 1–8, CIAT, Cali, Colombia, 1989.

25. A. V. Schoonhoven and O. Voysest, "Common beans in Latin America and their constraints," in Bean Production Problems in the Tropics, H. F. Schwartz and M. A. PastorCorrales, Eds., pp. 33–57, CIAT, Cali, Colombia, 1989.

26. S. Beebe and B. McClafferty, Biofortified Beans, HarvestPlus, CIAT, Cali, Colombia, 2006, http:// www.research4development.info/PDF/ Outputs/ Misc_Crop/beans.pdf.

27. R. Bressani, "Nutritive value of cowpea," in Cowpea: Research, Production and Utilization, S. R. Singh and K. O. Rachie, Eds., pp. 353–360, John Wiley & Sons, New York, NY, USA, 1985.

28. G. Lemma, W. Worku, and A. Woldemichael, "Moisture and planting density interactions affect productivity in cowpea (Vigna unguiculata)," Journal of Agronomy, vol. 8, no. 4, pp. 117–123, 2009.

29. B. B. Singh, "Recent genetic studies in cowpea," in Challenges and Opportunities for Enhancing Sustainable Cowpea Production, C. A. Fatokun, S. A. Tarawali, B. B. Singh, P. M. Kormawa, and M. Tamo, Eds., pp. 3–13, International Institute of Tropical Agriculture, Ibadan, Nigeria, 2002.

30. A. S. Langyintuo, J. Lowenberg-DeBoer, M. Faye et al., "Cowpea supply and demand in West and Central Africa," Field Crops Research, vol. 82, no. 2-3, pp. 215–231, 2003.

31. H. O. A. Elowad and A. E. Hall, "Influences of early and late nitrogen fertilization on yield and nitrogen fixation of cowpea under well-watered and dry field conditions," Field Crops Research, vol. 15, no. 3-4, pp. 229–244, 1987.

32. M. B. Kwapata and A. E. Hall, "Effects of moisture regime and phosphorus on mycorrhizal infection, nutrient uptake, and growth of cowpeas (Vigna unguiculata (L.) Walp.)," Field Crops Research, vol. 12, pp. 241–250, 1985.

33. R. L. Fery, "The cowpea: production, utilization, and research in the United States," Horticulture Review, vol. 12, pp. 197–222, 1990.

34. R. J. Carsky, B. Vanlauwe, and O. Lyasse, "Cowpea rotation as a resource management technology for cereal-based systems in the Savannas of West Africa," in Challenges and Opportunities for Enhancing Sustainable Cowpea Production, C. A. Fatokun, S. A. Tarawali, B. B. Singh, P. M. Kormawa, and M. Tamo, Eds., pp. 252–266, International Institute of Tropical Agriculture, Ibadan, Nigeria, 2002.

35. S. A. Tarawali, B. B. Singh, S. C. Gupta, et al., "Cowpea as a key factor for a new approach to integrated crop-livestock systems research in the dry savannas of West Africa," in Challenges and Opportunities for Enhancing Sustainable Cowpea Production, C. A.

Fatokun, S. A. Tarawali, B. B. Singh, P. M. Kormawa, and M. Tamo, Eds., pp. 233–251, International Institute of Tropical Agriculture, Ibadan, Nigeria, 2002.

36. N. Sanginga, K. E. Dashiell, J. Diels, et al., "Sustainable resource management coupled to resilient germplasm to provide new intensive cereal-grain-legume-livestock systems in the dry savanna,"Agriculture, Ecosystems & Environment, vol. 100, no. 2-3, pp. 305–314, 2003.

37. B. B. Singh, "Cowpea [Vigna unguiculata (L.) Walp," in Genetic Resources, Chromosome Engineering and Crop Improvement, R. J. Singh and P. P. Jauhar, Eds., vol. 1, pp. 117–162, CRC Press, Boca Raton, Fla, USA, 2005.

38. M. P. Timko, J. D. Ehlers, and P. A. Roberts, "Cowpea," in Genome Mapping and Molecular Breeding in Plants, C. Kole, Ed., vol. 3 of Pulses, Sugar and Tuber Crops, pp. 49–67, Springer, Berlin, Germany, 2007.

39. A. M. Murphy and P. E. Colucci, "A tropical forage solution to poor quality ruminant diets: a review ofLablab purpureus," Livestock Research for Rural Development, vol. 11, no. 2, 1999,

40. D. G. Cameron, "Tropical and subtropical pasture legumes," Queensland Agricultural Journal, vol. 114, no. 2, pp. 110–113, 1988.

41. R. L. Clem, "Animal production from legume-based ley pastures in southeastern Queensland," inTropical Legumes for Sustainable Farming Systems in Southern Africa and Australia, A. M. Whitbread and B. C. Pengelly, Eds., pp. 136–144, Australian Centre for International Agricultural Research, Canberra, Australia, 2004.

42. R. D. Armstrong, B. J. Kuskopf, G. Millar, A. M. Whitbread, and J. Standley, "Changes in soil chemical and physical properties following legumes and opportunity cropping on a cracking clay soil," Australian Journal of Experimental Agriculture, vol. 39, no. 4, pp. 445–456, 1999.

43. R. D. Armstrong, K. McCosker, S. B. Johnson et al., "Legume and opportunity cropping systems in central Queensland. 1. Legume growth, nitrogen fixation, and water use," Australian Journal of Agricultural Research, vol. 50, no. 6, pp. 909–924, 1999.

44. P. M. Maundu, G. W. Ngugi, and C. H. S. Kabuye, Traditional Food Plants of Kenya, National Museums of Kenya, English Press, Nairobi, Kenya, 1999.

45. C. V. Piper and W. J. Morse, The Bonavist, Lablab or Hyacinth Bean, Bulletin of the U.S. Department of Agriculture no. 318, USDA, Washington, DC, USA, 1915.

46. FAO Production Yearbook 2002, 56:83, 2002.

47. P. Zaidi, Drought Tolerance in Maize: Theoretical Considerations & Practical Implications, CIMMYT-Maize Program, Mexico DF, Mexico, 2002.

48. R. Çakir, "Effect of water stress at different development stages on vegetative and reproductive growth of corn," Field Crops Research, vol. 89, no. 1, pp. 1–16, 2004.

49. P. W. Heisey and G. O. Edmeades, "Maize production in drought-stressed environments: technical options and research resource allocation," World Maize Facts and Trends 1997/1998, 1999.

50. R. C. Muchow, "Comparative productivity of maize, sorghum and pearl millet in a semi-arid tropical environment II. Effect of water deficits," Field Crops Research, vol. 20, no. 3, pp. 207–219, 1989.

51. R. F. Grant, B. S. Jackson, J. R. Kiniry, and G. F. Arkin, "Water deficit timing effects on yield components in Maize," Agronomy Journal, vol. 81, no. 1, pp. 61–65, 1989.

52. J. Bolaños and G. O. Edmeades, "The importance of the anthesis-silking interval in breeding for drought tolerance in tropical maize," Field Crops Research, vol. 48, no. 1, pp. 65–80, 1996.

53. J. C. Vogel, A. Fuls, and A. Danin, "Geographical and environmental distribution of $C_3$ and $C_4$ grasses in the Sinai, Negev, and Judean deserts," Oecologia, vol. 70, no. 2, pp. 258–265, 1986.

# CHAPTER 2

# Effect of Urea, NPK and Compost on Growth and Yield of Soybean (Glycine max L.), in Semi-Arid Region of Sudan

**Samia Osman Yagoub,[1] Wigdan Mohamed Ali Ahmed,[2] and A. A. Mariod[3]**

1 Agronomy Department, College of Agricultural Studies, Sudan University of Science and Technology, P.O. Box 71 Khartoum North, Sudan

2 Training Department, Ministry of Agriculture, P.O. Box: 285 Khartoum, Sudan

3 Food Science and Technology Department, College of Agricultural Studies, Sudan University of Science and Technology, Khartoum North, Sudan

## ABSTRACT

A field experiment was conducted for two consecutive seasons (2009/2010 and 2010/2011) on the Demonstration Farm of the College of Agricultural Studies, Sudan University of Science and Technology at Shambat, to study the effect of some fertilizers on growth and yield of soybean (Glycine max L. merril). The experiment was laid out in a randomized complete block design (RCBD) with four replicates. The fertilizers treatments consisted of three types of fertilizers: urea

(180 kg/ha), NPK (361 kg/ha), compost (%) and the control. The results showed that fertilizers treatments in first season had significant difference on number of pods/plant, economic yield, harvest index. Mean while, highly significant difference on green, biological and straw yield. In second season fertilizers treatments had significant difference on plant height at 30 days, leaf area at 45 and 60 days, green yield, biological yield and straw yield.

## 1. INTRODUCTION

Soybean (Glycine max L.) is a member of Leguminosae family, rich in nutrients, and it is regarded as a nutrient storage. Soybean is not only seen as an oil plant but also used for various purposes [1]. Among grain legumes, soybean is an economically important crop that is grown in diverse environments throughout the world. Its adaptation to tropical and subtropical regions is still involving extensive breeding work [2]. To use land continuously for crop cultivation, incorporating organic and inorganic fertilizers to soil would provide multiple benefits for improving the chemical and physical status of the soil which results in improved crop yield [3]. Organic fertilizers include compost, farm yard manure (FYM), slurry, worm castings, urine, peat, green manure, dried blood, bone meal, fish meal, and feather meal [4]. Inorganic fertilizers include sodium nitrate, rock phosphate, limestone, ammonium nitrate, potassium nitrate, NPK fertilizers, muriate of potash (MOP), and supper phosphates [5]. Both organic and inorganic fertilizers are sources of mineral elements, which plants require for effective growth and development. Essential mineral elements are required in optimum amounts and are classified into micro and macro. Nitrogen, phosphorus, and potassium have great effects in plant growth and development. Their deficiencies or excesses result in marked effects on the growth and yield of crops. Nitrogen is a chlorophyll component, and it promotes vegetative growth and green colouration of foliage [6]. Phosphorus plays a major role in photosynthesis, respiration, energy storage, cell division, and maturation. Potassium is important in plant metabolism, protein synthesis, and chlorophyll development [7]. The most important crop nutrients in agricultural systems are nitrogen (N), phosphorus (P), and potassium (K) [8]. Most compound fertilizers will contain three elements essential for plant growth: NPK which stands for nitrogen (promotes leaf growth), phosphorus (root, flower, and fruit), and potassium (stem and root growth and protein analysis). Soybean nitrogen (N) requirements are met in a complex manner, as this crop is capable of utilizing both soil N (mostly in the form of nitrate) and atmospheric N (through symbiotic nitrogen fixation) [9]. The use of fertilizer is considered to be one of the most important factors to increase crop yield. Phosphorous has been shown to be an essential element, and its application has been shown to be important for growth, development, and yield of soybean [10]. Phosphorus deficiency is probably one of the greatest constraints for agriculture. Potassium often limits production and needs to be included in a soil fertility program; potassium should be included as corrective nutrient. Increases in soybean yield were obtained in response to K fertilization. The authors in [11] reported that potassium showed significant effect on yield and yield attributes of soybean in application of 40 kg ha. The increased growth of soybean may be due to optimum nutrient supply and better soil condition for growth of root and shoot of soybean crop [12]. Soybean was tried in several parts of Sudan since 1920s. It could be grown as a summer crop under irrigation in central and northern parts of the country or under rain-fed condition in the southern states [13]. Recently, the interest in soybean has been increased, and research work on soybean has been reactivated, this was due to increasing demand for soybean as an industrial crop in Sudan. The objective of this study was to investigate the effectiveness of urea, NPK, and compost on growth and yield of soybean under a semiarid region of Sudan conditions.

## 2. MATERIALS AND METHODS

Afield experiment was conducted for two consecutive seasons (2009/2010 and 2010/2011) in the Demonstrated Farm of the College of Agricultural Studies, Sudan University of Science and Technology, Shambat, Sudan, latitude 15°40″N, longitude 32°32″E, and altitude 386 m above sea level. The soil was montmorillonitic clay soil with a pH in the range of 7.8–8.5 (Table 1) [14].

**Table 1.** Full approximate analysis of soil in depth of 30 cm after sowing, season 2010/2011.

| Treatments | PH | Ece | Na meq/L | Ca+mg meq/L | K meq/L | $CO_2$ meq/L | $HCO_3$ meq/L | Cl meq/L | P ppm | $CaCO_3$ | N % | Sand % | Clay % | Silt % |
|---|---|---|---|---|---|---|---|---|---|---|---|---|---|---|
| Urea | 7.66 | 2.07 | 15.50 | 8.93 | 0.01 | 0.30 | 3.97 | 15.35 | 6.00 | 2.09 | 0.06 | 18 | 60.33 | 21.67 |
| Compost | 7.74 | 1.94 | 14.20 | 8.30 | 0.03 | 0.30 | 4.83 | 14.23 | 5.05 | 2.09 | 0.07 | 18 | 61.50 | 20.50 |
| NPK | 7.79 | 1.68 | 11.78 | 7.98 | 0.02 | 0.20 | 3.98 | 11.90 | 5.53 | 2.32 | 0.06 | 17 | 62.00 | 21.00 |
| Control | 7.74 | 2.05 | 14.90 | 9.10 | 0.02 | 0.35 | 5.35 | 14.78 | 4.58 | 2.31 | 0.06 | 17 | 62.25 | 20.75 |
| LSD | 0.26 | 0.62 | 7.38 | 2.52 | 0.01 | 0.28 | 2.41 | 6.47 | 1.94 | 0.98 | 0.01 | 1.92 | 4.09 | 2.82 |
| CV% | 2.10 | 19.93 | 32.74 | 18.34 | 42.36 | 60.98 | 33.19 | 28.75 | 22.89 | 27.91 | 14.11 | 6.87 | 4.15 | 8.39 |
| SE± | 0.16 | 0.39 | 4.61 | 1.57 | 0 | 0.17 | 1.50 | 4.04 | 1.21 | 0.61 | 0 | 1.20 | 2.55 | 1.76 |
| Mean before sowing | 7.91 | 1.29 | 10.88 | 4.62 | 0.01 | 0.02 | 0.8 | 10.61 | 2.4 | 1.64 | 0.05 | 18.3 | 61.8 | 19.9 |

* = significant, ** = highly significant, ns = nonsignificant, LSD = least significant difference, CV% = coefficient of variation, and SE± = standard error.

Climate of the locality has been described by [15] as semidesert and tropical with low relative humidity. The mean annual rainfall is about 160 mm, and the mean maximum temperature is more than 40°C in summer and around 20°C in cool season. Solar radiation is about 400–500 cal $cm^{-2}$ $day^{-1}$. The land was prepared by disc plough, disc harrowed, and leveled ridging up from north to south, and the spacing between ridges was 70 cm and between holes was 25 cm. The size of the plot was 3 × 3. 5 $m^2$ consisting of four ridges of 2.5 length. Sowing was done manually on the shoulder of the ridges, in the first season on the 16th of July 2009 and in the second season on the 8th of July 2010. Seeds were sown at a rate of 2-3 seeds per hole, and the fertilizers treatments were put immediately after sowing and then immediately irrigated. In both seasons, one genotype (1905 E) of Soybean was used. The treatments consist of three types of fertilizers urea: 46% N (180 kg/ha), NPK 23-23-0 (361 kg/ha), and compost and then control. The treatments were arranged in a randomized complete block design (RCBD) with four replications for two seasons.

Five plants were randomly selected and tagged in each plot and each treatment. Plants samples were taken at 30, 45, 60, and 75 days after sowing. Plant height (cm) was measured from the base of the main stem to the tip of the youngest leaf using measuring tape. The number of leaves/plant was taken by counting all leaves in the tagged plants, and the average number of leaves per plant was determined. Three leaves were randomly selected from tagged plants, and leaf area per plant was counted by using puncher. Days from sowing until first plant in each plot flowered, and days from sowing until 50% and 100% of plants in each plot flowered. The same tagged plants in each plot were harvested separated and immediately weighed, and the average weight of green plant was determined. Pods were separated from the plants, and dried normally by the sun. All pods in the tagged plants were counted and the average number of pods per plant was determined. Plants were dried in oven at 100°C for 24 hours and after that taken and immediately weighed, and the average weight of dry plant was determined. All these pods were threshed, and the seeds were taken and cleaned and then weighed in sensitive balance to determine seed yield per plant. All seeds were counted, and the average number of seeds per plant was determined. The number of seeds per pod was counted by using the following formula: number of seeds/pod = number of seeds/plant ÷

number of pods/plant. 100 seeds, were taken randomly from the tagged plants of each plot then weighed in sensitive balance to determine 100 seeds weight for each treatment. A meter length from the middle ridge of each plot was harvested and immediately weighed to determine green yield and then transferred to kg/ha. The same meter length was dried in oven at 100°C for 24 hours and after that taken and immediately weighed to determine biological yield, then it was transferred to kg/ha. Seeds in the same meter length from each plot were collected and dried naturally by the sun for a week, then cleaned and weighed in sensitive balance to determine seeds yield, and transferred to kg/ha. Straw yield was calculated by the following: straw yield = biological yield − economic yield (kg/ha). Harvest index was calculated by the following: harvest index = biological yield ÷ economic yield× 1 0 0. The data of plant parameters were analyzed yearwise on individual basis, and their means were computed. statistical analysis for ANOVA were carried out by using "MSTAT-C," whereas the means were compared through (LSD) tests least significant different [16] at $P = 0.05$ and excel program to illustrate and compare data on figures.

## 3. RESULTS AND DISCUSSION

The effect of fertilizers on plant height of soybean is presented in Table 2. Analysis of variance showed no significant difference between treatments in booth seasons at 30, 45, 60, and 75 days, except in the second season at 30 days. The highest mean plant height in the first season was 32.65 cm given by urea treatment; in the second season was 31.38 cm given by control. A similar result was found by [17, 18]. Meanwhile, [19] indicated that plant height decreased with increased urea. On the other hand, [11] reported that different doses of potassium and sulphur fertilizers had significant effect on the plant height of soybean. Nitrogen fertilization increased the mean plant height [20–23].

**Table 2.** Plant height of soybean genotype as affected by urea, NPK, and compost during 2009/10 and 2010/11 seasons.

| DAP | 2009/2010 | | | | 2010/2011 | | |
|---|---|---|---|---|---|---|---|
| treatments | 30 days | 45 days | 60 days | 75 days | 30 days | 45 days | 60 days |
| Urea | 17.53 | 24.20 | 27.80 | 32.65 | 15.35 | 20.88 | 26.38 |
| Compost | 17.80 | 23.40 | 27.40 | 32.00 | 18.23 | 23.85 | 29.43 |
| NPK | 16.25 | 18.80 | 29.15 | 30.40 | 15.83 | 23.05 | 30.20 |
| Control | 17.35 | 22.75 | 27.90 | 31.70 | 16.45 | 23.85 | 31.38 |
| LSD | 3.33 ns | 7.81 ns | 3.89 ns | 8.42 ns | 2.00* | 3.53 ns | 6.08 ns |
| CV% | 12.07 | 21.91 | 8.68 | 16.60 | 7.58 | 9.62 | 12.94 |
| SE± | 2.08 | 4.88 | 2.43 | 5.26 | 1.25 | 2.20 | 3.80 |

* = significant, ** = highly significant, ns = nonsignificant, LSD = least significant difference, CV% = coefficient of variation, SE± = standard error, and DAP = days after planting.

**Table 3.** Number of leaves/plant of soybean genotype as affected by urea, NPK, and compost during 2009/10 and 2010/11 seasons.

| DAP | 2009/2010 | | | | 2010/2011 | | |
|---|---|---|---|---|---|---|---|
| treatments | 30 days | 45 days | 60 days | 75 days | 30 days | 45 days | 60 days |
| Urea | 14.10 | 23.00 | 38.90 | 44.45 | 6.97 | 21.23 | 35.43 |
| Compost | 13.30 | 20.90 | 34.15 | 52.50 | 7.62 | 25.93 | 44.25 |
| NPK | 12.75 | 18.40 | 36.45 | 54.25 | 7.35 | 26.05 | 44.73 |
| Control | 12.10 | 21.90 | 38.25 | 54.55 | 7.30 | 19.58 | 31.85 |
| LSD | 2.64 ns | 7.44 ns | 8.20 ns | 13.45 ns | 0.96 ns | 7.23 ns | 14.12 ns |
| CV% | 12.64 | 22.10 | 13.89 | 16.35 | 8.22 | 19.48 | 22.59 |
| SE± | 1.65 | 4.65 | 5.13 | 8.41 | 0.60 | 4.52 | 8.83 |

* = significant, ** = highly significant, ns = nonsignificant, LSD = least significant difference, CV% = coefficient of variation, SE± = standard error, and DAP = days after planting.

The effect of fertilizers on the number of leaves per plant of soybean is presented in Table 3. Analysis of variance showeda no significant difference between treatments in both seasons at 30, 45, 60, and 75 days. Control showed slight increase in the number of leaves in the first season (54.55) and NPK (44.73) in the second season. A similar result was reported by [23, 24]. In on line with [22] that it was showed the difference in nitrogen levels had slight influence on the mean number of leaves per plant.

The effect of fertilizers on leaf area/plant of soybean is presented in Table 4. Analysis of variance showed no significant difference among treatments in first the season at all readings 30, 45, 60, and 75 days and showed significant difference among treatments in the second season at 45 and 60 days. The highest number of leafs per plant was in the first season 85.7 $cm^2$, in the second season was 135.46 $cm^2$ which was also given by control. Fertilizers had no effect in leaf area in booth seasons but nitrogen fertilizer increased leaf area in first season at 30 and 45 DAP. The authors in [25] found a different result.

The effect of fertilizers on the number of days to first, 50%, and 100% flowering of soybean is presented in Table5. Analysis of variance showed no significant difference between treatments. Days to flowering decreased with the increasing urea fertilization as reported by [21]. In addition, [22] observed that nitrogen fertilization can promote vegetative growth of soybean, and plants can be flowered around 33 days after sowing. Generally, high nitrogen levels of 60 and 120 kg/ha delayed flowering, whereas lower nitrogen levels had no influence on flowering.

The effect of fertilizers on the number of pods per plant, number of seeds per plant, and the number of seeds per pod of soybean is presented in Table 6. Analysis of variance showed no significant difference between treatments in both seasons except in number of pods per plant in the first season. The highest mean number of pods per plant in the first season was 95.95 given by compost treatment, and in the second season was 80.2 given by NPK treatment. The highest mean number of seeds per plant in the first season was 156.8 given by control, and in the second season it was 138.5 given by NPK treatment. Fertilizers increased the number of pods per plant. Similar results were reported by [22, 23, 26, 27]. In contrast, [19, 21] observed that increasing the levels of nitrogen fertilization had no effect on the mean number of pods per plant. The number of seeds per pod was slightly affected by nitrogen fertilization as noticed by [22, 23, 27].

**Table 4.** Leaf area per plant of soybean genotype as affected by urea, NPK, and compost during 2009/10 and 2010/11 seasons.

| DAP | 2009/2010 | | | | 2010/2011 | | |
|---|---|---|---|---|---|---|---|
| treatments | 30 days | 45 days | 60 days | 75 days | 30 days | 45 days | 60 days |
| Urea | 40.20 | 100.30 | 69.70 | 82.55 | 25.43 | 75.45 | 125.47 |
| Compost | 37.20 | 93.55 | 55.00 | 82.30 | 28.62 | 68.47 | 108.31 |
| NPK | 30.80 | 66.40 | 52.75 | 59.05 | 28.04 | 64.58 | 101.12 |
| Control | 38.30 | 112.60 | 64.80 | 85.70 | 30.33 | 82.89 | 135.46 |
| LSD | 16.22 ns | 56.50 ns | 16.58 ns | 41.68 ns | 6.40 ns | 11.17* | 24.80* |
| CV% | 27.69 | 37.90 | 17.12 | 33.66 | 14.23 | 9.59 | 13.18 |
| SE± | 10.14 | 35.32 | 10.37 | 26.06 | 4.00 | 6.98 | 15.50 |

* = significant, ** = highly significant, ns = nonsignificant, LSD = least significant difference, CV% = coefficient of variation, SE± = standard error, and DAP = days after planting.

**Table 5.** Number of days to first, 50%, and 100% flowering of soybean genotype as affected by urea, NPK, and compost during 2009/10 and 2010/11 seasons.

| Treatments | 2010/2011 | | |
|---|---|---|---|
| | Days to first flowering | Days to 50% flowering | Days to 100% flowering |
| Urea | 48.75 | 59.00 | 62.50 |
| Compost | 46.75 | 56.00 | 62.50 |
| NPK | 48.25 | 56.50 | 62.75 |
| Control | 47.75 | 57.50 | 63.00 |
| LSD | 2.65 ns | 2.59 ns | 3.04 ns |
| CV% | 3.46 | 2.82 | 3.03 |
| SE± | 1.66 | 1.62 | 1.62 |

* = significant, ** = highly significant, ns = nonsignificant, LSD = least significant difference, CV% = coefficient of variation, SE± = standard error.

**Table 6.** Number of pods per plant, number of seeds per plant, and number of seeds per pod of soybean genotype as affected by urea, NPK, and compost during 2009/10 and 2010/11 seasons.

| Treatments | 2009/2010 | | | 2010/2011 | | |
|---|---|---|---|---|---|---|
| | No of pods/plant | No of seeds/plant | No of seeds/pod | No of pods/plant | No of seeds/plant | No of seeds/pod |
| Urea | 80.30 | 139.10 | 1.71 | 64.40 | 114.20 | 1.75 |
| Compost | 95.95 | 150.75 | 1.59 | 75.05 | 123.70 | 1.65 |
| NPK | 82.55 | 124.65 | 1.53 | 80.20 | 138.50 | 1.73 |
| Control | 95.55 | 156.80 | 1.54 | 72.15 | 118.80 | 1.66 |
| LSD | 15.68* | 58.36 ns | 0.43 ns | 21.97 ns | 42.24 ns | 0.30 ns |
| CV% | 11.07 | 25.54 | 16.69 | 18.83 | 21.33 | 11.10 |
| SE± | 9.80 | 36.48 | 0.27 | 13.73 | 26.41 | 0.19 |

* = significant, ** = highly significant, ns = nonsignificant, LSD = least significant difference, CV% = coefficient of variation, SE± = standard error.

**Table 7.** Weight of seeds per plant and 100 seeds weight (g) of soybean genotype as affected by urea, NPK, and compost during 2009/10 and 2010/11 seasons.

| Treatments | 2009/2010 | | 2010/2011 | |
|---|---|---|---|---|
| | Weight of seeds/plant (g) | 100 seeds weight (g) | Weight of seeds/plant (g) | 100 seeds weight (g) |
| Urea | 10.81 | 9.45 | 6.82 | 6.04 |
| Compost | 11.42 | 9.78 | 7.46 | 5.73 |
| NPK | 9.68 | 9.40 | 8.34 | 6.47 |
| Control | 12.72 | 9.33 | 6.90 | 6.04 |
| LSD | 5.45 ns | 0.85 ns | 2.94 ns | 0.87 ns |
| CV% | 30.52 | 5.56 | 24.93 | 8.97 |
| SE± | 3.41 | 0.54 | 1.84 | 0.54 |

* = significant, ** = highly significant, ns = nonsignificant, LSD = least significant difference, CV% = coefficient of variation, SE± = standard error.

**Table 8.** Green yield and biological yield of soybean genotype as affected by urea, NPK, and compost during 2009/10 and 2010/11 seasons.

| Treatments | 2009/2010 | | 2010/2011 | |
|---|---|---|---|---|
| | Green yield | Biological yield | Green yield | Biological yield |
| Urea | 13285.98 | 7810.87 | 12928.83 | 6321.56 |
| Compost | 13954.60 | 8239.45 | 10714.50 | 5357.25 |
| NPK | 9953.77 | 6046.55 | 13964.57 | 6464.42 |
| Control | 15618.17 | 8953.75 | 10321.64 | 5107.25 |
| LSD | 2255.89** | 1313.84** | 3472.11* | 1233.10* |
| CV% | 10.68 | 10.58 | 18.12 | 13.26 |
| SE± | 1410.29 | 821.36 | 2170.63 | 770.89 |

* = significant, ** = highly significant, ns = nonsignificant, LSD = least significant difference, CV% = coefficient of variation, SE± = standard error.

The effect of fertilizers on the weight of seeds per plant and 100 seeds weight of soybean is presented in Table 7. Analysis of variance showed no significant difference among treatments in both seasons. The highest mean weight of seeds per plant in the first season was 12.72 (g) given by control, and in the second season it was 8.34 (g) given by NPK, and the highest mean of 100 seeds weight in the first season was 9.83 (g) given by control; in the second season was 6.47 (g) given by NPK treatment. Seeds weight is negatively correlated with the number of seeds per pod in many crops. In soybean, seed weight increased under nitrogen fertilization as indicated by [22, 23].

The effect of fertilizers on mean green and biological yield of soybean is presented in Table 8. Analysis of variance showed highly significant difference among treatments in the first season and significant difference, and in the second season. The highest mean green yield in the first season it was 15618.17 (kg/ha) given by control, and in the second season it was 13964.57 (kg/ha) given by NPK and the highest mean biological yield in first season was 8953.75 (kg/ha) given by control, in second season was 6464.42 (kg/ha) given by NPK treatment. The authors in [28] found similar result.

The effect of fertilizers on mean economic yield, straw yield, and harvest index of soybean is presented in Table9. In the first season, analysis of variance showed highly significant difference between treatments in straw yield, and significant difference in economic yield and harvest index. The highest mean straw yield was 6803.35 (kg/ha) given by control, economic yield was 2150.40 (kg/ha) given by control, and harvest index was 25.7% given by compost. In the second season, analysis of variance showed significant difference between treatments in straw yield and no significant difference in economic yield and harvest index although they were increased by fertilizers. The highest mean straw yield was 5310.68 (kg/ha) given by NPK, economic yield was 1193.63 (kg/ha) given by nitrogen, and harvest index was 19.86% given by compost treatment. These findings are in accordance with the results obtained by [11, 22, 26, 29–31]. The authors in [30] observed that the soybean grain yield increased with the application of 50 kg $K_2O$/ha. In contrast, [17, 19] indicated that high levels of nitrogen fertilization had no effect on seed yield of soybean.

**Table 9.** Economic yield, straw yield, and harvest index of soybean genotype as affected by urea, NPK, and compost during 2009/10 and 2010/11 seasons.

| Treatments | 2009/2010 | | | 2010/2011 | | |
|---|---|---|---|---|---|---|
| | Economic yield (g) | Straw yield (g) | Harvest index % | Economic yield (g) | Straw yield (g) | Harvest index % |
| Urea | 1929.33 | 5881.55 | 24.77 | 1193.63 | 5127.93 | 18.87 |
| Compost | 2118.61 | 6120.84 | 25.70 | 1049.13 | 4308.12 | 19.86 |
| NPK | 1227.17 | 4819.38 | 19.85 | 1153.74 | 5310.68 | 17.57 |
| Control | 2150.40 | 6803.35 | 23.29 | 819.80 | 4287.44 | 15.49 |
| LSD | 563.50* | 830.28** | 3.66* | 418.80 ns | 923.21 * | 5.30 ns |
| CV% | 18.98 | 8.79 | 9.78 | 24.84 | 12.13 | 18.44 |
| SE± | 352.28 | 519.06 | 2.9 | 261.82 | 577.16 | 3.31 |

* = significant, ** = highly significant, ns = nonsignificant, LSD = least significant difference, CV% = coefficient of variation, SE± = standard error.

# ACKNOWLEDGMENT

The authors would like to express they sincere appreciation and thanks for Sudan University of Science and Technology for financial support of this research.


# REFERENCES


1. B. Arslan, B. Yıldırım, and Ö. Dede, "Crop seed production and importance of plant oil requirement in Turkish economy," Yüzüncü Yıl Üniversitesi Ziraat Fakültesi Dergisi, vol. 3, no. 1-2, pp. 367–373, 1993.

2. ITC, Principle Oils and Seeds in World Trade. Atrader's Quide Volume II, International Trade Center, Geneva, Switzerland, 1990.

3. B. Basso and J. T. Ritchie, "Impact of compost, manure and inorganic fertilizer on nitrate leaching and yield for a 6-year maize-alfalfa rotation in Michigan," Agriculture, Ecosystems and Environment, vol. 108, no. 4, pp. 309–341, 2005.

4. R. J. Haynes and R. Naidu, "Influence of lime, fertilizer and manure applications on soil organic matter content and soil physical conditions: a review," Nutrient Cycling in Agroecosystems, vol. 51, no. 2, pp. 123–137, 1998.

5. M. D. Taylor, "Accumulation of cadmium derived from fertilisers in New Zealand soils," Science of the Total Environment, vol. 208, no. 1-2, pp. 64–68, 1997.

6. J. B. Jones, A guide for the Hydroponic and Soil-Less Culture Grower, Timber Press, Beaverton, Ore, USA, 1983.

7. S. U. Remison, Basic Principles of Crop Physiology, Sadoh Press Nig, Benin City, Benin, 2005.

8. V. O. Chude, W. B. Malgwi, I. V. Amapu, and O. A. Ano, Manual on Soil Fertilit Assessment, Federal Fertilizer Department (FFD) Incollaboration with FAO/National Special Programme for Food Security, Abuja, Nigeria, 2004.

9. M. Vera, N. Mrkovački, and M. Hrustić, "Interrelationship of nitrogen fixation potential and soybean yield. A Periodical of Scientific Research on Field and Vegetable," Crops, vol. 36, pp. 133–139, 2002.

10. K. M. Kakar, M. Tariq, F. H. Taj, and K. Nawab, "Phosphorous use efficiency of soybean as affected by phosphorous application and inoculation," Pakistan Journal of Agronomy, vol. 1, no. 1, pp. 49–50, 2002.

11. I. S. M. Farhad, M. N. Islam, S. Hoque, and M. S. I. Bhuiy, "Role of potassium and sulphur on the growth, yield and oil content of soybean (Glycine max L.)," An Academic Journal of Plant Sciences, vol. 3, no. 2, pp. 99–103, 2010.

12. S. D. Billore, K. Singh, A. P. Singh, and M. Bargale, "Energy productivity of sorghum-wheat sequence through long term use of chemical fertilizers," Indian Journal of Agronomy, vol. 39, no. 2, pp. 198–208, 1994.

13. M. O. Khidir, Oil Crops in the Sudan, Khartoum University Press, Khartoum, Sudan, 1997.

14. M. E. Abdel-Hafeez, Effect of partially acidulate phosphate rocks and triple superphosphate and their combinations on growth, mineral composition and yield of wheat, Ph.D. thesis, College of Agricultural Studies, University of Sudan for Science and Technology, Khartoum, Sudan, 2001.

15. H. S. Adam, The Agricultural Climate, Gezira University Press, Wad-Medani, Sudan, 2nd edition, 2002.

16. R. G. D. Steel, J. H. Torrie, and D. A. Dickeys, Principles and Procedures of Statistics. A Biometrical Approach, Mc Graw Hill Book, New York, NY, USA, 3rd edition, 1997.

17. M. M. Win, "Vegetable soybean yield response to different nitrogen rates," AVRDC-TOP 9 Training Report, Kasetsart University, Bangkok, Thailand, 1996.

18. S. A. Shehata and M. A. El-Helaly, "Effect of compost, humic acid and amino acid on yield of snap beans," Journal of Horticultural Science & Omamental Plants, vol. 2, no. 2, pp. 107–110, 2010.

19. C. J. Hantolo, "Nitrogen fertilizer effect on yield component of vegetable soybean," AVRDC-TOP 9 Training Report, Kasetsart University, Bangkok, Thailand, 1995.

20. M. F. Starling, C. W. Wood, and D. B. Weaver, "Starter nitrogen and growth habit effects on late-planted soybean," Agronomy Journal, vol. 90, no. 5, pp. 658–662, 1998.

21. L. Chuansong, "Nitrogen fertilizer effect on marketable yield of vegetable soybean," AVRDC-TOP 9 Training Report, Kasetsart University, Bangkok, Thailand, 1990.

22. T. Xuewen, "Effect of nitrogen fertilizer level on soybean yield," AVRDC-TOP 9 Training Report, Kasetsart University, Bangkok, Thailand, 1990.

23. G. A. Akbari, D. S. Scarisbrick, and W. T. Peat, "Soybean (Glycine max L. merrill) yield and yield components response to nitrogen supply and wither changes in South-East of England," Journal of Agriculture and Rural Development, vol. 3, no. 1, pp. 15–32, 2001.

24. P. N. Singh, A. S. Jeena, and G. R. Singh, "Effect of N.P. fertilizer on plant growth and root characteristic in soybean," Govind Ballabh Pant University of Agriculture, vol. 24, no. 2, pp. 127–129, 2001.

25. A. Polthanee and V. Tre-loges, "Effects of soaking seed and fertilizer placement on growth and yield of soybean grown after rice in the post-monsoon season in khon kaen province songklanakarin," Journal of Science and Technology, vol. 24, no. 2, pp. 197–207, 2002.

26. P. Tola, "Urea fertilizer and variety effect on vegetable soybean yield," AVRDC-TOP 9 Training Report, Kasetsart University, Bangkok, Thailand, 1995.

27. S. K. Agha, F. C. Oad, and U. A. Buriro, "Yield and yield components of inoculation and un-inoculation soybean under varying nitrogen levels Asian," Journal of Plant Sciences, vol. 3, no. 3, pp. 370–371, 2004.

28. P. Akbari, A. Ghalavand, A. M. Modarres Sanavy, and M. Agha Alikhani, "The effect of biofertilizers, nitrogen fertilizer and farmyard manure on grain yield and seed quality of sunflower (Helianthus annusL.)," Journal of Agricultural Technonogy, vol. 7, no. 1, pp. 173–184, 2011.

29. G. Seneviratne, L. H. J. Van Holm, and E. M. H. G. S. Ekanayake, "Agronomic benefits of rhizobial inoculant use over nitrogen fertilizer application in tropical soybean," Field Crops Research, vol. 68, no. 3, pp. 199–203, 2000.

30. A. R. Kheirelseed, Effect of inoculation N.P. on irrigated soybean in Gezira, M.S. thesis, University of Gezira, 1990.

31. H. S. Grewal, J. S. Kolar, and D. Singh, "Effect of split application of potassium on growth, yield and potassium accumulation by soybean(Glycine max L.) Merr.," Journal of Potassium Research, vol. 10, no. 2, pp. 173–177, 1994.

# CHAPTER 3

# Effects of Tree Shelters on the Survival and Growth ofArgania spinosa Seedlings in Mediterranean Arid Environment

**Chamchelmaarif Defaa,[1,2] Salwa Elantry,[3] Sanae Lahlimi El Alami,[3] Ahmed Achour,[1] Abdelhamid El Mousadik,[2] and Fouad Msanda[2]**

1 Direction Régionale des Eaux et Forêts et de la Lutte Contre la Désertification du Sud Ouest-Quartier Administratif, BP 520, 80000 Agadir, Morocco

2 Laboratoire de Biotechnologie et de Valorisation des Ressources Naturelles, FSA, Université Ibn Zohr, BP 8106, Dakhla, 80060 Agadir, Morocco

3 Centre de la Recherche Forestière, Avenue Omar Ibn Al Khattab, B.P. 763, Rabat, 10050 Agdal, Morocco

## ABSTRACT

The argan tree is endemic species of Morocco. It occupies an area of more than 8700 km$^2$ and plays essential ecological and economical roles. In spite of their value, the argan woodlands are subject to rapid and uncontrolled degradation during the last decades, mainly due to overgrazing and systematic collection of argan nuts. The present study was carried out to investigate the effects of two types of tree shelters on survival and growth of Argania spinosa seedlings planted in the southwest of Morocco in order to improve the results of reforestation programs which usually end by repeated failures. The experiment was conducted in the Mesguina forest for two growing seasons after transplantation in the field. After two years, the use of tree shelters significantly increased tree survival and allowed a gain of 20%. Height growth was positively affected by tree shelters. The use of tree shelters showed no significant decrease in basal diameter. In contrast, the height to diameter ratios of sheltered trees were much higher compared to the control. Thus, the use of the tree shelters could aid the early establishment and growth of Argania spinosa under conditions similar to those of the experiment.

## 1. INTRODUCTION

The argan forests (Argania spinosa (L.) Skeels) extend over 8700 square kilometers [1]. For centuries, argan tree has shaped the socioeconomic life of the southwestern Morocco, becoming a flagship species for the region and for the country [2]. This space has been declared by UNESCO's Man and the Biosphere (MAB) Programme as biosphere reserve in 1998. The argan tree is the major tree species of Macaronesian formations in a climate characterized by very large occult precipitation [3]. Argan tree is a thermophilic and xerophytic tree species, evolving in warm temperate arid bioclimate (along the coast and in the plains) and in warm temperate semiarid bioclimate (High Atlas and Anti-Atlas), with annual rainfall ranging from 400 mm in the north (Safi), 250 mm in the Souss Valley, and 150 mm in the Anti-Atlas to less than 100 mm in the southern areas of desert nature [4]. This tree, which can live up to 250 years [5], provides multiple uses for the local population: its very hard wood and the shell of the fruit are used for heating, the leaves and the pulp of its fruit and the oil cake are a valuable fodder for the flock, its thorny branches are used as fence for agricultural plots, edible oil and cosmetics are derived from its seed, and finally the rich soil of the forest area is a very favorable area for intercropping.

During recent years, the argan tree faces many constraints that result in the weakening and degradation of natural forest ecosystems. The argan forests face, in recent decades, the changes in climate and many anthropozoogenic disturbances that result in a reduction in area and density and inexorably lead to the weakening and deterioration of the natural forest ecosystems. As part of the global forest resources assessment [6], official figures provided by Morocco and relating to the degradation of the argan tree show a decrease of 111 $km^2$ of the total area of the argan tree between 1990 and 2005, which corresponds to a decrease of 7.4 $km^2$/year. On the other hand, the argan forests are in almost total lack of natural regeneration [7].

With the aim of preserving this natural resource and to reverse this decline, the Forest Department opted for assisted regeneration in the late 90s and led a large reforestation program. However, successful argan plantations are hard to establish, and many reforestation projects have resulted in repeated failures, depriving local users of a vital space for their flocks for too long periods. Some breeders do not respect the exclosure despite the implementation of a state compensation of 350 Dh/ha ($ 3600/$km^2$) in favor of users' associations. Infringement of the exclosure by flocks in these areas is one of the main factors of failure of forest regeneration [8]. Added to that is damage by wild animals (hares, gazelles, and boar) and rodents (Barbary ground squirrel and voles) that multiply significantly in these closed spaces where phytomass quickly becomes very important. Under these conditions, individual protection of young seedlings, using tree shelters, is required. Being developed to protect trees from animal damage, these plastic tubes showed their ability to improve early growth by creating a "greenhouse effect" around each tree [9]. Numerous scientific publications have been devoted to the use of shelters in reforestation for several woody species, highlighting many benefits [9–13] but also some of their disadvantages [14–16]. No work has been devoted to the study of the effect of tree shelters on young argan seedlings in the field conditions. The present work aims to test how these techniques can improve the regeneration of argan tree by assessing the effects of individual protection seedlings on its growth parameters under field conditions for the first two growing seasons after transplantation.

## 2. MATERIALS AND METHODS

### 2.1. Study Site

The experimental study was carried out at the Mesguina argan forest (9°23′W, 30°25′N, elevation 115 m) in southwestern Morocco, 25 km east of Agadir city. The study site is located in a fenced area and strictly closed to grazing. The climate is Mediterranean with an annual mean rainfall of 232.8 mm. The annual mean temperature is 20.0°C.

Soil samples were collected with auger at depths of 0–20 cm and 20–40 cm. These samples were analyzed by standard methods to determine the granulometric and physicochemical characteristics. The soil of the study site is sandy clay, with neutral pH, and poor in organic matter (Table 1).

**Table 1.** Physicochemical characteristics of the plot study soil.

| Soil depth (cm) | C (%) | N (‰) | C/N | OM (%) | pH (1:5) | P (mg/kg) |
|---|---|---|---|---|---|---|
| 0–20 | 2.12 | 0.85 | 24.89 | 3.66 | 7.17 | 36.17 |
| 20–40 | 0.32 | 0.08 | 40.62 | 0.56 | 7.36 | 36.25 |

### 2.2. Materials

The plantation was established in mid-March 2010. The seeds used in this study had the same origin, and the harvest was carried out in argan forest of Ida Ou Moumen region. Argan seedlings were grown in the Oulad Teima forest nursery located 20 km from study area. This nursery produces hundreds of thousands of argan seedlings per year and constitutes the main supplier for the argan tree regeneration programs in the whole region. The seedlings were grown in 500 cm$^3$ and 7 cm square containers filled with 75–25% (v/v) white peat and forest soil and transplanted in the field after 11 months in the nursery. These rigid containers are made of black polypropylene. They have an inverted truncated pyramidal shape with an open base in order to promote root air-pruning. The tree shelter used is a double-walled light green translucent polypropylene vented tube. Openings at the base of the shelter provide ventilation inside the tube. Two heights of the shelters were tested: 120 cm and 150 cm.

## 2.3. Experimental Design and Treatments

A complete randomized block design with three blocks was used. Each block contained 102 argan seedlings: 34 with 120 cm shelters, 34 with 150 cm shelters, and 34 argan seedlings as control (without shelters). On the plot, seedlings were planted with a density of 204 seedlings/ha, in cubic holes with dimensions of 0.60 m  0.60 m in surface and 0.60 m in depth. A catchment area of 20 cm high and 1 m around the plant was established. The 204 shelters (sizes 120 and 150 cm) were then placed around the treated plants and 102 argan seedlings were left as control. Until August, every month a regular watering of 30 L per plant was provided during the first year. During the second growing season, no watering was provided to plants.

For each plant and throughout the duration of the study, plant height and basal diameter were measured. In addition to measurements at the planting date, two measurements were made during the first year (October 30 and December 26) and one measurement (October 30) was made for the following year, except basal diameter for which the measurement was performed only during the first year. Indeed, during the second year, the development of plants and their ramifications in the shelters did not allow measuring the extent of the basal diameter.

## 2.4. Statistical Analysis

Quantitative variables (survival, growth height, basal diameter, and height to diameter ratio) were analyzed with repeated measures. During the growing seasons 2010 and 2011, 3 measurements were made. The initial values of each variable measured just after planting were considered as a covariable. An analysis of covariance (ANCOVA) was then conducted.

The comparison of averages was made using Scheffé test calculated at [17]. Previously, data had undergone logarithmic transformations to satisfy the assumptions required for the covariance analysis. Data for plants survival had undergone an angular transformation arcsine type of the square root (), the raw data being expressed by values from 0 to 1 [18].

# 3. RESULTS AND DISCUSSION

## 3.1. Survival

After two years of experimentation the results showed that the survival rate was 94.12% for sheltered trees versus 74.51% for unsheltered trees. This represents a very high rate that very few regeneration schemes have reached in the region.

Statistical treatments showed highly significant intergroup differences both for the first year () and for the second year () for the treatments on survival of argan seedlings. Results showed a positive effect of tree shelters on survival of argan seedlings (Figure 1).

The tree shelters improved significantly the survival rates (). The difference with the control was about 20% in both the first and the second year. Survival rates for seedlings under 120 cm shelters as well as 150 cm were not significantly different. After two years, survival rate of 94% in sheltered plants can be considered exceptional in view of the aridity in this region.

The analysis of Figure 1 shows that annual deaths that occurred in unsheltered seedlings are more important than in sheltered ones. Almost all mortalities were observed during the first year. These results confirm the previous findings [13, 19–24] which showed that mortality was higher for unprotected plants compared to sheltered plants. Similar results were found in arid climate for Eucalyptus tree [12]. In continental climate, the use of shelters allowed a mean survival gain of 17% for cottonwood and Siberian elm [25]. Indeed, the shelters played likely a decisive role around the seedlings by creating a favorable microclimate by reducing light intensity [26, 27] and plant transpiration [28] during the summer. According to some studies [29], the soil temperature around unsheltered seedlings may exceed the soil temperature registered in tree shelters by 10 to 20°C. Moreover, the shelter protected the plants against attack of many mammals in the fenced plot: hares and Barbary ground squirrel (Atlantoxerus getulus) that attacks seedling's shoots and especially its apical part which greatly hinders in growth and causes the death of the plant in case of repeated assaults on young argan trees. Contrastingly, some authors have shown no effect on survival [16, 30, 31].

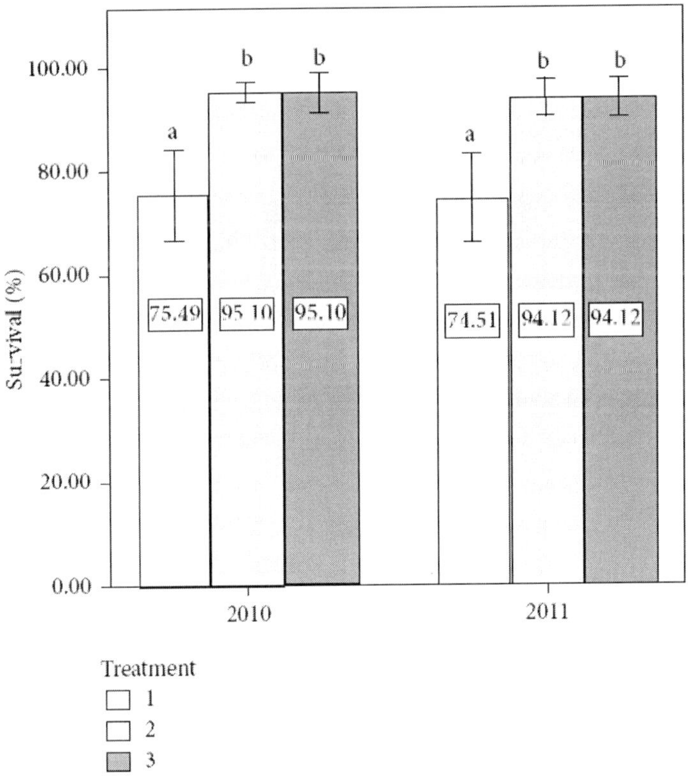


**Figure 1.** Survival rate (%) of argan seedlings for 2010 and 2011. Treatment 1: control. Treatment 2: seedlings with 120 cm shelters. Treatment 3: seedlings with 150 cm shelters. Values sharing the same letter within the same year are not significantly different at 5% level (Scheffé test).

Rainfall and temperatures recorded during the two campaigns may explain much of the mortality rates throughout the plot. Thus, the most important mortalities were observed during the first year regardless of the treatment, due to the difficult acclimatization of the argan seedlings. This acclimatization was made all the more difficult by the fact that the field conditions, especially during the 2009/2010 season, have been unfavorable: rainfall has not exceeded 56.2 mm after the planting date and temperatures were above the annual average. Several heat waves have hit the study area between April and September. These temperatures reached extreme values of 47.3°C and 48.3°C, respectively, on August 12 and August 26, 2010. The shelters gave to the seedlings an additional source of protection from the desiccating wind and thus less evapotranspiration losses. Significant environmental benefits arise due to saved irrigation water by the use of tree shelters in establishing trees in semiarid areas [13]. Thus the use of water and nutrients was better managed and utilized by seedlings that were in the shelter [25].

During the second year, which was very wet, the argan trees found favorable conditions for growth and mortalities greatly diminished. The results are consistent with the findings of studies over several tree species [29, 32]. Moreover, the first year is crucial to overcoming the phenomenon of transplanting stress. This stress is resulting from the confinement of root to the planting hole, poor root-soil contact, and low root permeability [33].

## 3.2. Height Growth

The statistical treatment showed highly significant effects in the first year and very highly significant effects in the second year on the height growth of argan seedlings (Table 2).

Shelters have significantly stimulated the height growth of argan seedlings compared to the control during the two years of experimentation (Figure 2).

**Table 2.** Effects of treatments on height growth.

| Date | Df | *F* | *P* |
|---|---|---|---|
| October 2010 | 2 | 5.629 | 0.004 |
| December 2010 | 2 | 6.432 | 0.002 |
| October 2011 | 2 | 18.783 | <0.001 |

Df: degree of freedom.

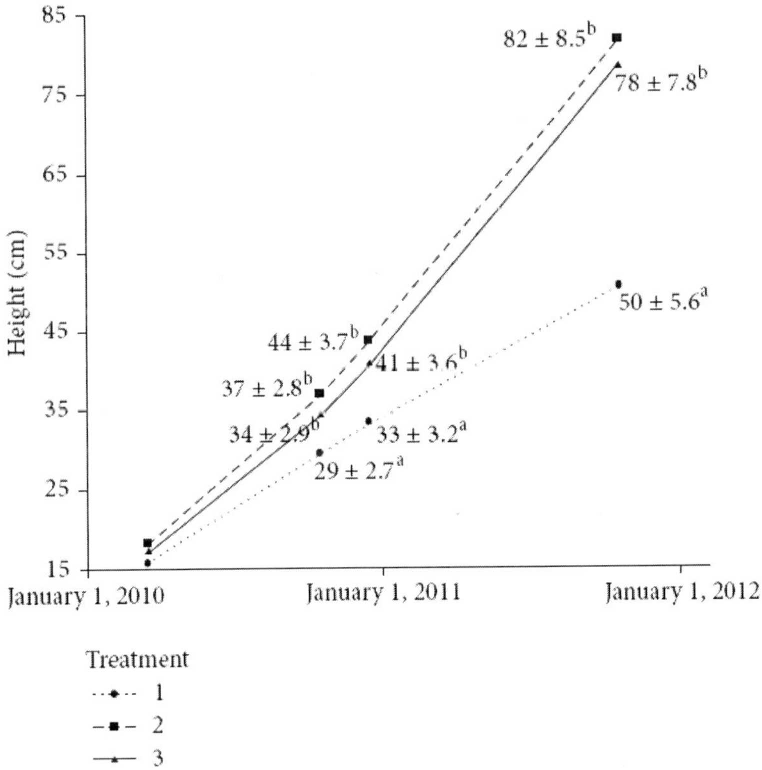


**Figure 2.** Growth height according to tree shelter factor. Treatment 1: control. Treatment 2: seedlings with 120 cm shelters. Treatment 3: seedlings with 150 cm shelters. Values sharing the same letter within same date are not significantly different at 5% by the Scheffé test.

The highest increases were observed for argan seedlings under 120 cm tubes. There was no significant difference compared to the 150 cm shelters. These classic effects of stimulation of height growth of seedlings have been reported by many authors for many tree species and climates [19, 20, 23, 24, 26, 34–38]. Similar results were obtained for oak trees in the south of England [9], for walnut in the south of France [14], and for the argan tree after one growing season under nonlimiting water conditions [39]. On the other hand, tree shelter did not improve the height growth of Casuarina cristata and Eucalyptus moluccana [30].

During the first year, the growth rates of sheltered argan seedlings were higher compared to the control. The same trend was observed during the second year when watering was not applied, with much sharper dynamics in sheltered trees. This growth's kinetics were stimulated by favorable water conditions: additional watering during the first year and the exceptional rainfall during the 2nd year.

## 3.3. Basal Diameter

During the first growing season, the measurements showed that the average of basal diameters had become two times higher than its initial value for all treatments (Figure 3). The greatest diameters were observed for the control, however without significant difference with the sheltered trees.

Thus, for control seedlings, the basal diameter increased from 2.91 mm at planting date to 6.98 mm after 9 months, which corresponds to 140% increase (Figure 4). The shelter seedlings had the lowest mean basal diameter. With the exception of publications on two oak species [9] where the tubes allowed good growth of the diameter and the lack of effect on magnolia and oak [34], the results are consistent with those found for several tree species [14–16] and also for Argania spinosa under nonlimiting conditions [39].

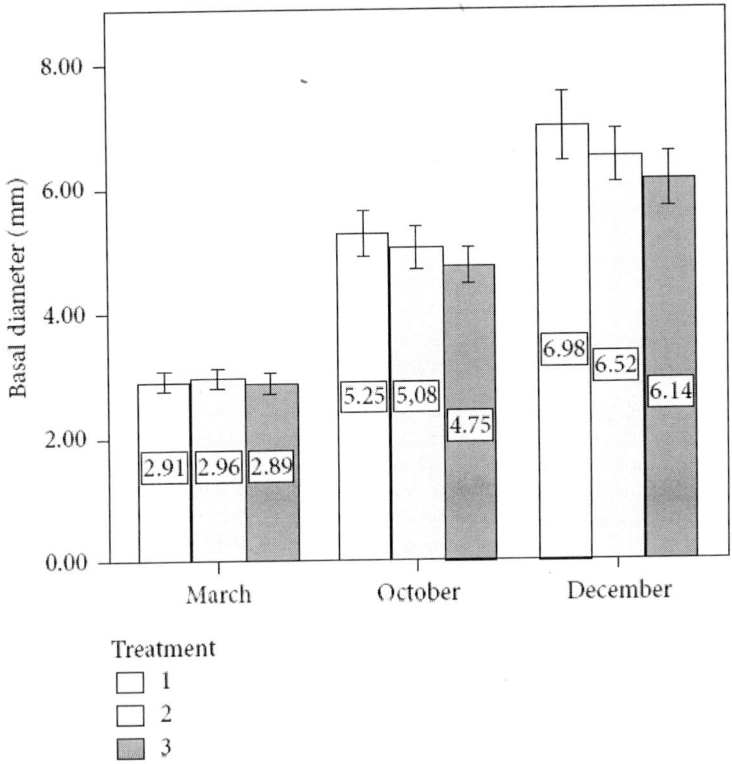


**Figure 3.** Mean basal diameters at planting date (March 2010) and after 7 months (October 2010) and 9 months (December 2010). Treatment 1: control. Treatment 2: seedlings with 120 cm shelters. Treatment 3: seedlings with 150 cm shelters.

## 3.4. Height to Diameter (H/D) Ratio

The results showed a highly significant effect of treatments on the H/D ratio () both for October and December 2010.

The height to diameter ratio (H/D ratio) is an important factor, which describes the type of stem that each tree develops [40]. A lower ratio means a more vigorous tree [39].

At seven months as well as nine months, the H/D ratio of the control was statistically much lower than that for the sheltered trees (Table 3).

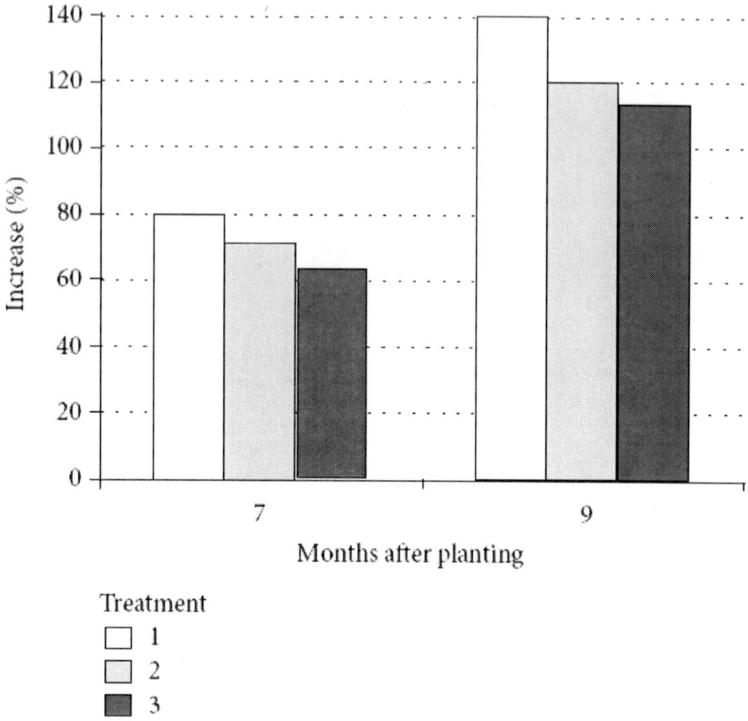


**Figure 4.** Increases of basal diameters for the argan tree seedlings. Treatment 1: control. Treatment 2: seedlings with 120 cm shelters. Treatment 3: seedlings with 150 cm shelters.

**Table 3.** ratio mean of argan seedlings for October and December 2010. Values sharing the same letter within the same date are not significantly different at 5% by Scheffé test.

| Date | Treatment | Mean | SD |
|---|---|---|---|
| | Control | $56.31^{a}$ | 14.17 |
| October 2010 | T120 | $72.82^{b}$ | 16.06 |
| | T150 | $72.96^{b}$ | 21.16 |
| | Control | $47.88^{a}$ | 12.42 |
| December 2010 | T120 | $66.68^{b}$ | 16.36 |
| | T150 | $66.82^{b}$ | 21.96 |

A low H/D ratio is the sign of a stable tree with vigorous and powerful roots [40]. It has also been reported that shelter promotes rapid and vigorous root growth immediately after planting for Quercus ilex and Pinus halepensis seedlings under nonlimiting water availability conditions in Mediterranean climate [41].

On the other hand, the height to diameter ratio decreased with the length of the tube, but with no significant difference. These results show that H/D ratio has decreased between October and December. This decrease generally results in more vigorous trees. These results confirm those found for argan trees under nonlimiting water conditions [39], where the H/D ratio had a value of 53 for the control versus 74 to 104 for the sheltered trees depending on the type of shelter.

# 4. CONCLUSIONS

Two years after planting, the results have shown the beneficial effect of shelters on the survival rate of argan seedlings with a gain of 26%.

Tree shelters have, also, stimulated the height growth of the argan trees. Thus, the argan seedlings began to emerge from their shelters one year after planting. Seven months after planting, growing height showed a gain of 28% and 64%, respectively, for the first and second year for sheltered trees (120 cm). This result is very important as it may reduce the duration of deferred grazing. Diameter growth basal was not significantly affected by the introduction of tree shelters. Tree shelters (120 cm) induced a ratio H/D 69% higher compared to the control but which can still be considered acceptable.

Survival rate is the decisive parameter for the success or failure of a regeneration project. The use of tree shelter showed survival rates of 94% after two years. This can be considered exceptional on harsh conditions in southwestern Morocco with irregular rainfall and extreme summer temperatures that sometimes exceeded 48°C. The use of tree shelters should be considered one of the most effective ecotechnological treatments to enhance both seedling survival and growth [23].

With regard to the height of tree shelters, the results showed no significant differences for the studied parameters. Therefore, the use of 120 cm tree shelters would be suitable since it would allow a faster emergence of the plants from their shelters and a better H/D ratio.

This study is one of the first field trials with Argania spinosa in its natural distribution area. Two growing seasons after planting, tree shelter, designed initially to protect against browsing, could enhance the establishment and growth of argan trees planted on harsh conditions in southwestern Morocco.

## ACKNOWLEDGMENTS

The authors would like to thank the Forestry Research Center (CRF) who supported this study. The authors are grateful to the Regional Department of Water and Forests (DREFLCDSO) and their staff for their assistance in all phases of this research, including seedling supply and plantation installation. Special thanks are given to Tubex for providing tree shelters.

## REFERENCES

1. Haut Commissariat aux Eaux et Forêts et à la Lutte Contre la Désertification, “Les formations forestières,” HCEFLCD, 1996,

2. L. Kenny, “Histoire de l'arganier,” in Atlas de l'Arganier et de l'Arganeraie, L. Kenny and I. De Zborowski, Eds., pp. 12–38, Institut Agronomique et Vétérinaire Hassan II, Rabat, Morocco, 2007.

3. A. Benabid, “Biodiversité écosystémique et paysagère,” in Flore et écosystèmes du Maroc: Evaluation et Préservation de la Biodiversité, A. Benabid, Ed., pp. 225–286, Ibis Press, Paris, France, 2000.

4. F. Msanda, A. El Aboudi, and J. P. Peltier, “Biodiversité et biogéographie de l'arganeraie marocaine,”Cahiers Agricultures, vol. 14, no. 4, pp. 357–364, 2005.

5. P. Boudy, “Monographie et traitement de l'arganier,” in Economie Forestière Nord-Africaine Tome deuxième, P. Boudy, Ed., pp. 382–416, Larose, Paris, France, 1950.

6. Food and Agriculture Organization of the United Nations, Evaluation des Ressources Forestières Mondiales 2010: Rapport National—Maroc, FAO, 2010

7. H. Faouzi and J. Martin, “Soutenabilité de l'arganeraie marocaine,” Confins, vol. 20, 2014

8. C. Defaa, A. Achour, A. Hossayni, R. Bellefontaine, A. El Mousadik, and F. Msanda, “Analyse de l'itinéraire technique d'un périmètre exceptionnellement réussi de régénération d'arganier,” in Actes du 1er Congrès International sur l'Arganier, pp. 83–92, Institut National de la Recherche Agronomique, Agadir, Morocco, December 2013.

9. G. Tuley, “The growth of young oak trees in shelters,” Forestry, vol. 58, no. 2, pp. 181–195, 1985.

10. G. Armand, “Techniques de protection contre les cervidés,” Revue Forestière Française, vol. 44, pp. 99–103, 1992.

11. J. H. McAdam, “An evaluation of tree protection methods against Scottish Blackface sheep in an upland agroforestry system,” Forest Ecology and Management, vol. 45, no. 1–4, pp. 119–125, 1991.

12. D. Gautier, T. Tapsou, and D. Christian, “Quand l'éleveur élève aussi...des arbres. Un essai de protection avec des manchons de jeunes arbres en milieu soudano-sahélien fortement pâturé,” in Savanes Africaines: Des Espaces en Mutation, des Acteurs Face à de Nouveaux Défis, pp. 27–31, Cirad, Cirad, Montpellier, France, 2007.

13. J. C. Arnold and S. M. Alston, "Life cycle assessment of the production and use of polypropylene tree shelters," Journal of Environmental Management, vol. 94, no. 1, pp. 1–12, 2012.

14. C. Dupraz, "Les protections de plants à effet de serre: ce qu'en pensent les arbres," Revue Forestière Française, vol. 49, pp. 417–432, 1997.

15. D. W. Burger, G. W. Forister, and R. Gross, "Short and long-term effects of tree shelters on the root and stem growth of ornamental trees," Journal of Arboriculture, vol. 23, no. 2, pp. 49–56, 1997.

16. M. N. Jiménez, F. B. Navarro, M. Á. Ripoll, I. Bocio, and E. De Simón, "Effect of shelter tubes on establishment and growth of Juniperus thurifera L. (Cupressaceae) seedlings in Mediterranean semi-arid environment," Annals of Forest Science, vol. 62, no. 7, pp. 717–725, 2005.

17. H. Scheffé, "A method for judging all contrasts in the analysis of variance," Biometrika, vol. 40, pp. 87–104, 1953.

18. P. Dagnelie, "Inférence statistique à une et à deux dimensions," in Statistique Théorique et Appliquée Tome 2, P. Dagnelie, Ed., De Boeck, Brussels, Belgium, 2011.

19. J. D. Bakker, L. B. Colasurdo, and J. R. Evans, "Enhancing Garry oak seedling performance in a semiarid environment," Northwest Science, vol. 86, no. 4, pp. 300–309, 2012.

20. J. G. Pausas, C. Bladé, A. Valdecantos et al., "Pines and oaks in the restoration of Mediterranean landscapes of Spain: new perspectives for an old practice—a review," Plant Ecology, vol. 171, no. 1-2, pp. 209–220, 2004.

21. H. Chaar, T. Mechergui, A. Khouaja, and H. Abid, "Effects of treeshelters and polyethylene mulch sheets on survival and growth of cork oak (Quercus suber L.) seedlings planted in northwestern Tunisia," Forest Ecology and Management, vol. 256, no. 4, pp. 722–731, 2008.

22. F. M. Padilla, J. D. D. Miranda, R. Ortega, M. Hervás, J. Sánchez, and F. I. Pugnaire, "Does shelter enhance early seedling survival in dry environments? A test with eight Mediterranean species," Applied Vegetation Science, vol. 14, no. 1, pp. 31–39, 2011.

23. J. Piñeiro, F. T. Maestre, L. Bartolomé, and A. Valdecantos, "Ecotechnology as a tool for restoring degraded drylands: a meta-analysis of field experiments," Ecological Engineering, vol. 61, pp. 133–144, 2013.

24. C. Ceacero, R. Navarro-Cerrillo, J. Díaz-Hernández, and A. Del Campo, "Is tree shelter protection an effective complement to weed competition management in improving the morpho-physiological response of holm oak planted seedlings?" iForest, vol. 7, no. 5, pp. 289–299, 2014.

25. A. M. Stiffarm, Phytoremediation case study [M.Phil. thesis], Kansas State University, Manhattan, Kan, USA, 2015.

26. J. Bellot, J. M. Ortiz De Urbina, A. Bonet, and J. R. Sánchez, "The effects of treeshelters on the growth ofQuercus cocciferaL. seedlings in a semiarid environment," Forestry, vol. 75, no. 1, pp. 89–106, 2002.

27. F. Famiani, P. Proietti, M. Micheli et al., "Effects of tree shelters on young olive (Olea europaea) tree growth and physiology," New Zealand Journal of Crop and Horticultural Science, vol. 35, no. 3, pp. 303–312, 2007.

28. J.-E. Bergez and C. Dupraz, "Transpiration rate of Prunus avium L. seedlings inside an unventilated treeshelter," Forest Ecology and Management, vol. 97, no. 3, pp. 255–264, 1997.

29. M. F. Urretavizcaya and G. E. Defossé, "Effects of nurse shrubs and tree shelters on the survival and growth of two Austrocedrus chilensis seedling types in a forest restoration trial in semiarid Patagonia, Argentina," Annals of Forest Science, vol. 70, no. 1, pp. 21–30, 2013.

30. G. M. Dunn, M. S. Cant, and M. R. Nester, “Potential of two tree shelters to aid the early establishment and growth of three Australian tree species on the Darling Downs, south-east Queensland,” Australian Forestry, vol. 57, no. 3, pp. 95–97, 1994.

31. T. Mechergui, B. Hasnaoui, M. Pardos, and N. Boussaidi, “First-year effects of tree shelters and mulching on survival and growth of zeen oak (Quercus canariensis Lamk.) seedlings planted in North-Western Tunisia,” Revue D'Écologie, vol. 67, no. 1, pp. 3–18, 2012.

32. D. De Steven, “Experiments on mechanisms of tree establishment in old-field succession: seedling survival and growth,” Ecology, vol. 72, no. 3, pp. 1076–1088, 1991.

33. A. N. Burdett, “Physiological processes in plantation establishment and the development of specifications for forest planting stock,” Canadian Journal of Forest Research, vol. 20, no. 4, pp. 415–427, 1990.

34. D. W. Burger, P. Svihra, and R. Harris, “Tree shelter use in producing container-grown trees,”HortScience, vol. 27, no. 1, pp. 30–32, 1992.

35. P. Balandier, J. L. Guitton, and H. Rapey, “Amélioration des tubes abris protégeant les jeunes arbres contre les animaux,” Ingénieries, no. 4, pp. 41–48, 1995.

36. J. A. Oliet and D. F. Jacobs, “Microclimatic conditions and plant morpho-physiological development within a tree shelter environment during establishment of Quercus ilex seedlings,” Agricultural and Forest Meteorology, vol. 144, no. 1-2, pp. 58–72, 2007.

37. T. Mechergui, M. Pardos, N. Boussaidi, B. Hasnaoui, and D. F. Jacobs, “Development of cork oak (Quercus suber L.) seedlings in response to tree shelters and mulching in northwestern Tunisia,” Journal of Forestry Research, vol. 24, no. 2, pp. 193–204, 2013.

38. F. Ponder Jr. and J. W. Van Sambeek, “Nine-year performance of four hardwoods on a harvested site with and without fertilizer tree shelters, and weed mats in Southern Illinois,” in Proceedings of the

18th Central Hardwoods Forest Conference, General Technical Report GTR-NRS, pp. 63–71, US Forest Service, USDA, Newtown, Pa, USA, July 2013.

39. M. Fallah, C. Dupraz, and M. Dauzat, "Impact des protections individuelles à effet de serre sur des plants d'arganier en conditions hydriques non limitantes," Forêt Méditerranéenne, vol. 3, pp. 235–240, 2001.

40. J. de Champs, "Mesures sylvicoles préventives," Revue Forestière Française, vol. 4, pp. 313–322, 1987.

41. A. V. De Castro, J. A. Oliet, J. Puértolas, and D. F. Jacobs, "Light transmissivity of tube shelters affects root growth and biomass allocation of Quercus ilex L. and Pinus halepensis Mill," Annals of Forest Science, vol. 71, no. 1, pp. 91–99, 2014.

# CHAPTER 4

# Seed Germination Characteristics of Rhus tripartitum(Ucria) Grande and Ziziphus lotus (L.): Effects of Water Stress

**Zouaoui Refka,[1,2] Ksontini Mustapha,[1] and Ferchichi Ali[2]**

[1] Ecology and Sylvo-Pastoral Improvement Laboratory, National Research Institute of Rural Engineering, Water and Forests (INRGREF), Tunisia

2 Arid and Oases Cropping Laboratory, Arid Area Institute (IRA), 4119 Medenine, Tunisia

## ABSTRACT

Ziziphus lotus (L.) Lam. (Rhamnaceae) and Rhus tripartitum or Sumac (Anacardiaceae) are two indigenous species from arid and semiarid regions of Tunisia, characterized by a severe climate where dry seasons are very long. The combined action of anthropogenic factors and climate in arid regions caused a gradual threat of plant assets. In this context, an experimental study of the effects of water stress (0 to −1 MPa) on seeds has identified the water requirements germinal stage of both species. The results showed that both species were able to germinate at relatively low water potentials. However, beyond −0.6 MPa, germination was completely inhibited for R. tripartitum, when it reached for another 50% for Z. lotus. Increasing the concentration of $PEG_{6000}$progressively inhibited germination in both species. Only Z. lotus could be considered tolerant of water stress, because, to −1 MPa, seeds germinated with a rate of 17%. It resulted in that the species Z. lotus presented an adaptive capacity to aridity much greater than that observed for R. tripartitum.

# 1. INTRODUCTION

Drought stress is considered to be the main environmental factor limiting plant growth and yield of many agronomic and horticultural crops, especially in semi-arid areas. In Mediterranean-type ecosystems, seasonal water shortage is the main factor constraining survival and growth of plants.

Soil depth and texture are considered the most important edaphic properties that influence the moisture regime in arid environments with episodic rainfall. To date, a great deal of effort has been focused on physiological process underlying plant responses to drought stress.

The current situation of arid and desert areas of Tunisia (three quarters of the area) is in rapid decline of natural vegetation cover associated with an erosion of biodiversity [1, 2]. This decline is attributed to particular stressful environmental conditions, land clearing, and overgrazing, resulting in effects of increasingly adverse ecological (desertification) and economical.

In this area, the steppic vegetation is dominated by tall perennial grasses [3]. But, sparse trees and shrubs are among the most threatened species because of their excessive use for domestic purposes and their poor regeneration performances [4, 5]. On the other hand, under arid bioclimate, the flora has been subjected to a high, permanent increase of human pressure since at least the last century. Such a situation contributed and might have induced the phenomenon of desertification [5, 6].

In Tunisia, several attempts have been made to restore degraded rangelands [7–9]. The improvement of the methods based on seed dispersal and seed sowing would enable to know more about seed response to main environment factors especially drought stress in arid land. This factor could influence germination parameters [10, 11]. Under these conditions, seed germination monitoring in relation to water is very important to determine the colonisation capacity of species [12].

In this area, several native species are potentially interesting under aspects of dune stabilization and extension of plant cover, including some Ziziphus species and Rhus tripartitum. In addition, they improve the stability of ecosystems where they are present, contribute to reduce the risks of desertification, and are helpful in restoring degraded ecosystems [6, 13–16].

These shrubs are also a feeding resource for livestock especially during the summer, when the alternative herbaceous species have wilted [17]. According to Barroso et al. [18], these species relieve sand movement. Currently, Z. lotus L. and R. tripartitum are rarefied in this area under overgrazing and their artificial reintroduction is necessary for the restoration of degraded ecosystems.

Accurate knowledge on the germination requirements of Z. lotus L. and R. tripartitum species is now required for successful uses of these two species in operations of artificial regeneration in Tunisia. Therefore, the effects of low water availability, simulated by $PEG_{6000}$, on seed germination were studied for these two tree species located in arid and semi-arid areas of Tunisia.

# 2. MATERIALS AND METHODS

## 2.1. Site Collection Seed

The seeds of both species of Ziziphus lotus and Rhus tripartitum were collected from native shrubs (or low trees) in the National Park of Bou-Hedma (34°39′ N and 9°48′ E) (Figure 1) in September 2009. This park is located in the Governorate of Sidi Bouzid (central southern Tunisia). The bioclimate in this site is Mediterranean arid, with temperate winters and with large interannual and interseasonal variations of precipitation.

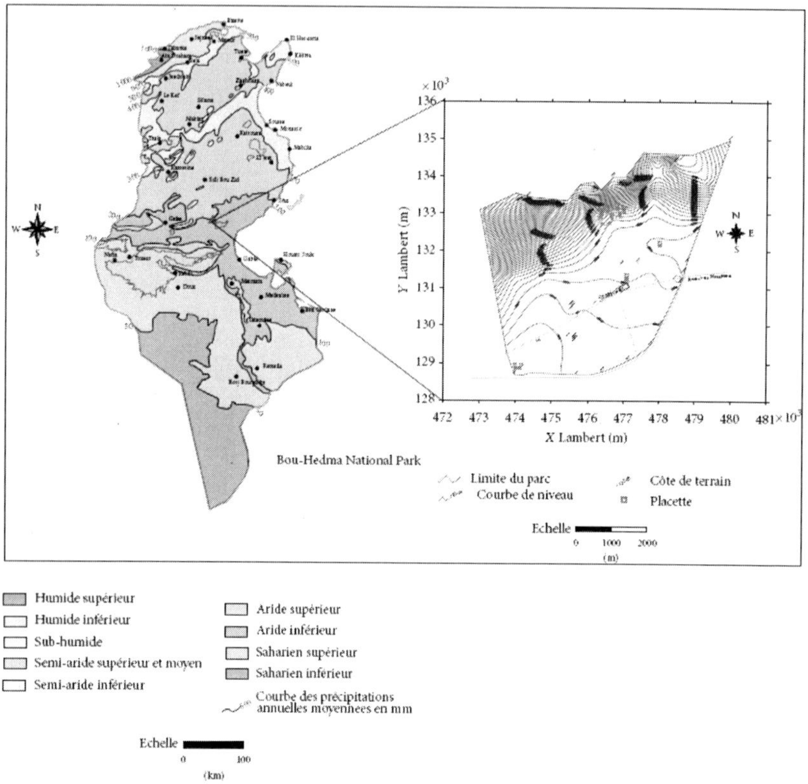


**Figure 1.** Location of the Bou-Hedma National Park in the bioclimatic map of Tunisia [32].

The mean annual average of rainfall is 180 mm in the plain and 250 mm on the crest of the mountain. Mean temperature ranges between 3.9°C throughout the coldest month (January) and 36.2°C throughout the hottest month (August).

## 2.2. Germination Experiments

Drought stress was induced by polyethylene glycol ($PEG_{6000}$) treatments. Six levels of osmotic potential ($\Psi\pi$): 0, −0.2, −0.4, −0.6, −0.8, and −1.0 MPa were tested and prepared from the formula developed by Mickel and Kaufman (1973) in [19]:

$$\Psi h = -\left(1.18 \cdot 10^{-2}\right) * C - \left(1.18 \cdot 10^{-4}\right) * C^2 + \left(2.67 \cdot 10^{-4}\right) * CT + \left(8.39 \cdot 10^{-7}\right) * C^2 T, \quad (1)$$

where is the concentration of $PEG_{6000}$ in $g \cdot L^{-1}$ of water, is the temperature in °C, and is osmotic potential in bars. Distilled water served as a control. Seeds of the same form for both species (Z. lotus and R. tripartitum) have been selected and sterilized by aqueous solution of 75% ethanol for 5 min to prevent fungal attack and rinsed in several changes of sterile distilled water. The seeds were germinated in 9 cm sterile Petri dishes lined with two sterile Whatman number 1 filter papers with 5 mL of distilled water or the respective test solutions [20]; there were 20 seeds per Petri dish and five replicates in each treatment. Germination tests were conducted in complete darkness at 30°C. Distilled water was added into the Petri dishes each day to maintain the concentration of the test solutions. Seeds were considered to be germinated when the radicle emergenced. After seed sowing, the germinated seeds were counted and eliminated every second day for 21 days.

## 2.3. Methods of Germination Expression

The characteristics of seedling emergence were (FG), number determined: final germination percentage of days to first germination (delay of germination), critical limits of germination and mean time to germinate (MTG).

1) Final germination percentage (FG) was calculated as the cumulative number of germinated seeds with normal radicles [19]. , where is the number of seeds that had germinated at each counting.

2) Mean Time of germination: MTG was estimated according to the formula:

$$MTG = \frac{\Sigma (ni \times di)}{N}, \quad (2)$$

where *n* is the number of seeds germinated at day , the incubation period in days, and the total number of germinated seeds in the treatment [21].

### 2.4. Statistical Analysis

Statistical analysis was performed using two-way analysis of variance (ANOVA and DUNCAN) to test the effects of water stress, species, and their interactions, on germination characteristics. All statistical methods were performed using SPSS, version 16.

## 3. RESULTS

### 3.1. Influence of Water Stress on the Kinetics of Germination

The kinetics of germination by osmotic stress conditions is presented in Figure 2. It reflects the sensitivity of the species to water stress. In control condition, the kinetic curves show three phases of germination: latency, accelerating exponentially, and finally step corresponding to a stop germination after attaining the maximum germination. The depressive effect of water stress on germination occurs during one or all of these three phases, depending on the degree of the lowering of water potential and species studied.

It results in a slower processing speed visible from −0.2 MPa, and thereafter increases. Whose, the delay of germination the most extensive (3 days) is obtained to −1 MPa for Z. lotus while 7 days to −0.6 and −0.8 MPa for R. tripartitum.

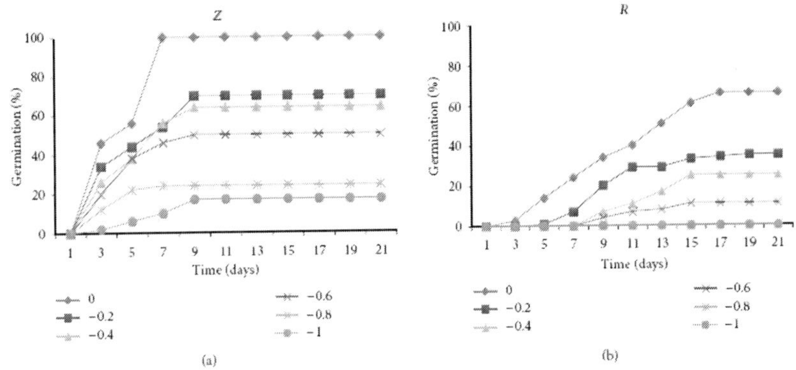


**Figure 2.** Cumulative germination percentage of Ziziphus lotus (Z) and Rhus tripartitum (Z) seeds at different $PEG_{6000}$ solutions of different osmotic potentials (0 to −1 MPa) (Z).

## 3.2. Final Germination Percentage

Final germination percentage of Z. lotus seeds were recorded in controls with (100%) and only 66% of R. tripartitum. The germination capacity is greatly affected by water stress (Figure 3), analysis of variance showed that there is a highly significant difference between treatments and species (Z) (Table 1).

Test ANOVA with two factors, calculated at 5% level, presented a treatment effect and species very highly significant () and the interaction between them (Table 1).

Duncan test (Table 2) discriminates two homogeneous groups under species effect: G1 = A to Z. lotus and G2 = B to R. tripartitum.

Duncan's expresses homogeneous groups according to the order of decreasing water potential (Table 3). It is found that upon application of −0.2 MPa, the germination capacity is significantly reduced especially for R. tripartitum where it exceeds 50%.

**Table 1.** Results of a two-way ANOVA showing effect of water stress on the germination rate in both species: Ziziphus lotus and Rhus tripartitum.

| Source of variation | ddl | Mean square | *F* ratio | *P* |
|---|---|---|---|---|
| Species | 1 | 2699,51 | 304,46 | 0,000*** |
| Treatment | 5 | 92,87 | 10,47 | 0,000*** |
| Interaction Sp*Trt | 5 | 89,79 | 10,13 | 0,000*** |
| Error | 48 | 8,87 | | |
| Total | 59 | | | |

Numbers representing the value of *F*. ***$P < 0.0001$ effect very highly significant.

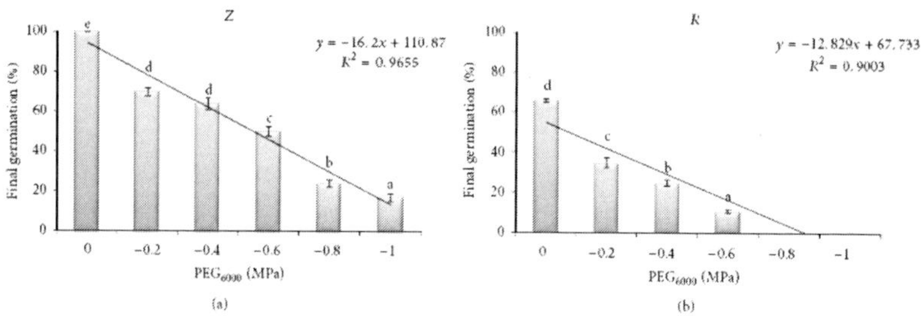


**Figure 3.** Variation of the final germination percentages (%) of Ziziphus lotus (Z) and Rhus tripartitum (R) seeds at different $PEG_{6000}$ solutions of different osmotic potentials (0 to −1 MPa) (N= 5).

**Table 2.** Comparison of the speed of germination by Duncan test of two species ofZiziphus lotus and Rhus tripartitum as a result of water stress (species effect).

| Species | Mean | Group |
|---|---|---|
| *Z. lotus* | 20,03 | A |
| *R. tripartitum* | 6,62 | B |

Different letters indicate significant differences between species.

**Table 3.** Mean comparison (Duncan test ) of final percentage and mean time to germination of Ziziphus lotus (Z) and Rhus tripartitum (R) seeds under osmotic stress condition.

| | | Germination characteristics | | | |
|---|---|---|---|---|---|
| | Water potentiel (MPa) | Final germination percentage (FG) | Decrease (%) | CV (%) | MTG (days) |
| *Ziziphus lotus* | 0 | $100 \pm 0^{f}$ | 0 | $20,35 \pm 2,16^{a}$ | $4,96 \pm 0.55$ |
| | −0.2 | $70 \pm 2^{d}$ | 30 | $19,27 \pm 2,63^{a}$ | $5,27 \pm 0,74$ |
| | −0.4 | $64 \pm 3,03^{d}$ | 36 | $19,15 \pm 3,91^{a}$ | $5,4 \pm 1,12$ |
| | −0.6 | $50 \pm 2,44^{c}$ | 50 | $21,66 \pm 5,19^{a}$ | $4,85 \pm 1,22$ |
| | −0.8 | $24 \pm 1,78^{b}$ | 76 | $24,28 \pm 5,49^{a}$ | $4,27 \pm 0,83$ |
| | −1 | $17 \pm 1,94^{a}$ | 83 | $15,48 \pm 4,29^{a}$ | $6,87 \pm 1,87$ |
| *Rhus tripartitum* | 0 | $66 \pm 0,84^{a}$ | 34 | $10,40 \pm 0,85^{b}$ | $9,67 \pm 0,83^{b}$ |
| | −0.2 | $35 \pm 2,45^{b}$ | 65 | $12,05 \pm 1,02^{a}$ | $8,35 \pm 0,70^{c}$ |
| | −0.4 | $25 \pm 1,58^{c}$ | 75 | $8,24 \pm 0,8^{c}$ | $12,22 \pm 1,14^{a}$ |
| | −0.6 | $11 \pm 0,84^{d}$ | 89 | $9,01 \pm 1,28^{c}$ | $11,27 \pm 1,46^{a}$ |
| | −0.8 | $0^{e}$ | 100 | $0^{d}$ | $0^{d}$ |
| | −1 | $0^{e}$ | 100 | $0^{d}$ | $0^{d}$ |

The values followed by the same letter are not significantly different at $P > 0.05$ each value is a mean of five replicates of 20 seeds.
CV: kinetics of germination; FG: final germination percentage; MTG: mean time to germinate.

### 3.3. Critical Limits of Germination

Critical limits of germination of both species were highly different. Seed germination of Z. lotus vanishes up there from −1 MPa but at −0.8 MPa for R. tripartitum. So, the critical limit of germination of Z. lotus are the order of −1 MPa while it was −0.6 MPa for R. tripartitum (Figure 3).

### 3.4. Influence of Water Stress on the Final Percentage and Mean Germination Time (MTG)

To better understand the effect of water stress on germination of Z. lotus and R. tripartitum, we showed the relationship between mean germination time and the final percentage. In fact, in Z. lotus, a water potential of −0.2 MPa causes a decrease in germination capacity of around 30%. This decrease becomes important to −0.8 MPa (76%) and even more accentuated at −1 MPa (83%).

Whereas R. tripartitum, a potential of −0.2 MPa, causes a 65% reduction of germination capacity, this reduction increases (89%) at −0.6 MPa and reached 100% at −0.8 MPa.

The results obtained for GMT of Z. lotus range from ≈5 to 7 days respectively, for 0 and −1 MPa, whereas Rhus tripartitum, varies from 9.67 to 11 days, respectively, to 0 and −0.6 MPa. So, the lowering of the water potential leads to an increase in the mean germination time.

The comparison of both species responding to water stress shows that Z. lotus is the most effective (tolerant) for most treatments compared to R. tripartitum (Figures 4 and 5).

## 4. DISCUSSION

The main objective of our paper was to assess the germination characteristics of Z. lotus and R. tripartitumunder abiotic constraints. The most important result showed that germination capacity of studied species under the environmental constraints and factors is sufficiently assured to consider these species for a reforestation program and the extend consequently their area of distribution. Additionally, studied species responded differently to stress induced by $PEG_{6000}$. However, Z. lotus manifested better adaptation to drought conditions compared to R. tripartitum.

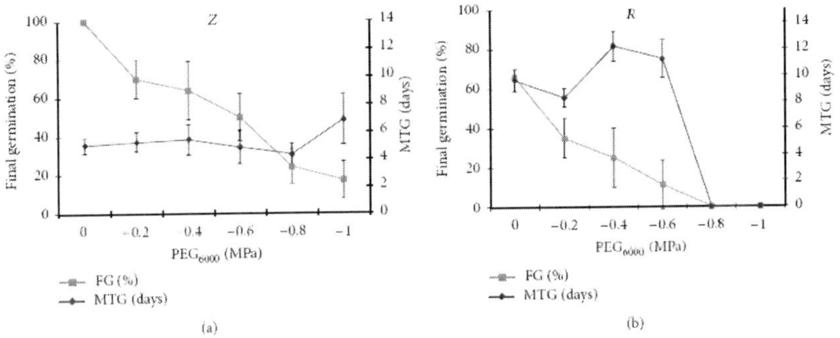


**Figure 4.** Variation of the final germination percentages (%) and the mean time to germination (MTG, days) of Ziziphus lotus (Z) and Rhus tripartitum (R) seeds (n=5) at different $PEG_{6000}$ solutions of different osmotic potentials (0 to −1 MPa).

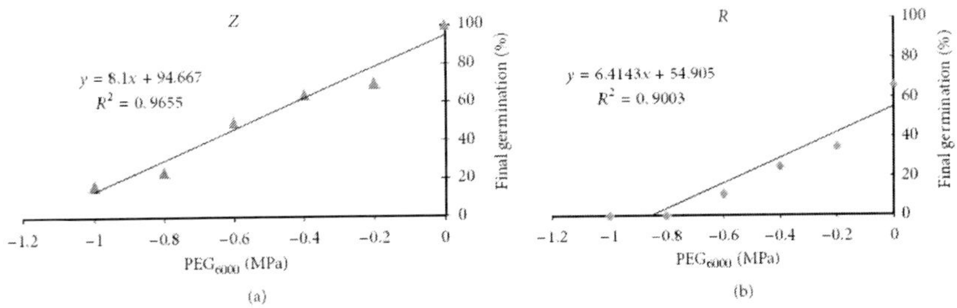


**Figure 5.** Regression plots for mean final germination percentages of Ziziphus lotus (Z) and Rhus tripartitum (R) seeds at different $PEG_{6000}$ concentrations. Values (n=100) are from the six treatments with five replicates.

Under arid bioclimate, the successful establishment of plants largely depends on the success of germination. The success of shrubs under warm and dry conditions is primarily dependent on optimal conditions (25/35°C and 12 h night/12 h day) for germination and recruitment [22].

Our results suggested that germination of Z. lotus seeds continued despite the increase of osmotic potential (Table 1). Up to −0.6 MPa, germination rate was superior to 50% for Z. lotus but completely inhibited for R. tripartitum (Table 1). The higher germination percentage of Z. lotus compared to R. tripartitum may be related to its better adaptation to water deficits. Our results are consistent with the results of Ne'eman et al. [23] who showed that germination of Rhus coriaria was reduced by 80% at water potential of −0.26 MPa. In the same way, Noumi et al. [24] showed that up to −0.7 MPa seeds germination of R. tripartitum is totally inhibited.

Under controlled conditions, the highest germination percentage of studied species was obtained for non-stressed seeds. The variation of $PEG_{6000}$ concentration has significantly affected germination rate.

Overall, our results showed that the increase of water stress introduces the decrease in germination rate of studied species. This finding is consistent with the results of Ne'eman et al. [23] who showed that germination ofRhus coriaria was reduced by 80% at water potential of −0.26 MPa. In the same way, Chuang et al. [25] showed that the critical value of Periploca sepium was −1 MPa and the maximum value was −1.4 MPa under the PEG. This finding corroborate with results from Lotus creticus ssp eucreticus, Plantago albicans L. Subsp.albicans.L.Hand Rhanterium suaveolens Desf. subsp. Suaveolens (Desf) [26].

In contrast to our results, the critical threshold of germination of Z. lotus and R. tripartitum is very highest compared to others species growing under arid and Saharan climates such as Lotus creticus (−0.7 MPa), Plantago albicans (−0.5 MPa), and Rantherium suaveolens (−0.3 MPa) [27]. In the same context, Sharma [28] showed that germination of Atriplex vesicaria Heward and Atriplex nummularia Lindl was totally inhibited up to −0.2 and −0.4 MPa, respectively.

In Tunisia, several attempts have been made to restore degraded rangelands [7–9, 13, 14]. One important technique for the rehabilitation of degraded ecosystems is to improve the dispersal of seeds, which is labour saving when compared to transplanting nursery seedlings [15]. Arid zones in Tunisia, which cover more than 70% of the total area [2], are characterized by drought and high temperature and more or less general salinity. In addition many shrub species such as Z. lotus, R. tripartita, Periploca angustifolia, Retama raetam, Acacia raddiana, Ceratonia siliqua, and Lycium shawii are valuable species for afforestation in arid area of Tunisia as they withstand the harsh climatic conditions, improve the stability of ecosystems where they are present and contribute to reduce the risks of desertification, and are helpful in restoring degraded ecosystems [6, 13–16].

In the field, establishment of Z. lotus and R. tripartitum varies according to the season (cool/warm) and the geographical and ecological distribution as well as the amount of rainfall. Seasonal fluctuations of rain in deserts of Tunisia explain the variable establishment pattern of this plant during the course of a year, governed by seasonally different temperatures and the rainfall events. From the present results, it is evident that germination percentage decreases with increasing water stress and comes to the limit at −1 MPa.

# 5. CONCLUSION

Some demographic studies [29, 30] suggest that the establishment of many woody species from arid and semi-arid areas is a rare event, which is mainly due to the occurrence of excessive water stress caused by low rainfall and/or high evapotranspiration rates. However, the establishment of these woody species would be possible only after years of above normal and well-distributed rainfall [31].

In conclusion, Z. lotus and R. tripartitum species manifested different levels of adaptation to drought. Increasing the concentration of $PEG_{6000}$ progressively inhibited germination for both species. Only Z. lotus can be considered tolerant to water stress, since, to −1 MPa, seeds germinate at a rate of 17%. As a result, Z. lotuspresents an adaptive capacity to aridity much larger than that observed for R. tripartitum. Therefore, these findings form the basis for future trials involving the use of indigenous shrubs in the restoration of rangelands.

## ACKNOWLEDGMENTS

We thank Lazher Hamdi Manager of the National Park of Bou-Hedma for his help.

## REFERENCES

1. H. N. Le Houerou, “La végétation de la Tunisie steppique (1) (Structure, écologie, sociologie, répartition, évolution, utilisation, biomasse, productivité) (avec référence aux végétations analogues d’Algérie, de Libye et du Maroc),” Annales de l’Institut National de la Recherche Agronomique de la Tunisie, vol. 42, no. 5, 622 pages, 1969.

2. C. Floret and R. Pontanier, L’aridité en Tunisie présharienne, climat, sol, vegetation et aménagement [Ph.D. thesis], University of Science and Technology of Languedoc, Montpellier, France, 1982.

3. H. N. Le Houérou, “Biogeography of the arid steppeland north of the Sahara,” Journal of Arid Environments, vol. 48, no. 2, pp. 103–128, 2001.

4. S. Ashkenazi, Acacia Trees in the Negev and the Arava, Israel: A Review Following Reported Large-Scale Mortality, vol. 17, Hakeren HaKayemet LeIsrael, Jerusalem, Palestine, 1995.

5. A. Hanafi and S. Jauffret, “Are long-term vegetation dynamics useful in monitoring and assessing desertification processes in the

arid steppe, southern Tunisia," Journal of Arid Environments, vol. 72, no. 4, pp. 557–572, 2008.

6. J. P. Gupta, R. K. Aggarwal, and N. P. Raikhy, "Soil erosion by wind from bare sandy plains in western Rajasthan, India," Journal of Arid Environments, vol. 4, no. 1, pp. 15–20, 1981.

7. E. Le Floc'h, M. Neffati, M. Chaieb, and R. Pontanier, "Un essai de réhabilitation en zone aride. Le cas de Menzel Habib (Tunisie)," in L'Homme Peut-il Refaire ce qu'il a Défait ? R. Pontanier, Ed., pp. 139–160, John Libbey Eurotext, Paris, France, 1995.

8. S. Jauffret and M. Visser, "Assigning life-history traits to plant species to better qualify arid land degradation in Presaharian Tunisia," Journal of Arid Environments, vol. 55, no. 1, pp. 1–28, 2003.

9. A. Ouled Belgacem, M. Neffati, V. P. Papanastasis, and M. Chaieb, "Effects of seed age and seeding depth on growth of Stipa lagascae R. & Sch. seedlings," Journal of Arid Environments, vol. 65, no. 4, pp. 682–687, 2006.

10. T. Tlig, M. Gorai, and M. Neffati, "Germination responses of Diplotaxis harra to temperature and salinity," Flora, vol. 203, no. 5, pp. 421–428, 2008.

11. Z. Noumi, Acacia tortilis (Forssk.) Hayne subsp. raddiana (Savi) Brenan en Tunisie pré- saharienne: structure du peuplement, réponses et effets biologiques et environnementaux [Ph.D. thesis], University of Bordeaux 1 and Science Faculty of Sfax, Tunisia, 2010.

12. I. A. Ungar, "Germination ecology of halophytes," in Contribution To the Ecology of Halophytes, D. N. Sen and K. S. Rajpurohit, Eds., pp. 143–154, The Hague, 1982.

13. J. Aronson, C. Floret, E. Le Floc'h, C. Ovalle, and R. Pontanier, "Restoration and rehabilitation of degraded ecosystems in arid and semi-arid lands. I. A view from the South," Restoration Ecology, vol. 1, no. 1, pp. 8–17, 1993.

14. J. Aronson, C. Floret, E. Le Floc'h, C. Ovalle, and R. Pontanier, "Restoration and rehabilitation of degraded ecosystems in arid and semi-arid lands. II—case studies in southern Tunisia, central Chile, and northern Cameroon," Ecological Restoration, vol. 1, pp. 168–187, 1993.

15. F. T. Maestre, S. Bautista, J. Cortina, and J. Bellot, "Potential for using facilitation by grasses to establish shrubs on a semiarid degraded steppe," Ecological Applications, vol. 11, no. 6, pp. 1641–1655, 2001.

16. J. Castro, R. Zamora, J. A. Hódar, and J. M. Gómez, "Use of shrubs as nurse plants: a new technique for reforestation in Mediterranean Mountains," Restoration Ecology, vol. 10, no. 2, pp. 297–305, 2002.

17. V. P. Papanastasis, P. D. Platis, and O. Dini-Papanastasi, "Effects of age and frequency of cutting on productivity of Mediterranean deciduous fodder tree and shrub plantations," Forest Ecology and Management, vol. 110, no. 1–3, pp. 283–292, 1998.

18. F. G. Barroso, T. F. Martinez, T. Paz, C. L. Alados, and J. Escós, "Relationship of Periploca laevigata(Asclepidaceae) tannins to livestock herbivory," Journal of Arid Environments, vol. 53, no. 1, pp. 125–135, 2003.

19. M. Gorai, T. Tlig, and M. Neffati, "Influence of water stress on seed germination characterestics in invasive Diplotaxis harra (Forssk) boiss (Brassicaceae) in arid zone of Tunisia," Journal of Phytology, vol. 1, no. 4, pp. 249–254, 2009.

20. M. Rejili, A. M. Vadel, A. Guetet et al., "Influence of temperature and salinity on the germination ofLotus creticus (L.) from the arid land of Tunisia," African Journal of Ecology, vol. 48, no. 2, pp. 329–337, 2010.

21. J. L. Brenchley and R. J. Probert, "Seed germination responses to some environmental factors in the seagrass Zostera capricorni from eastern Australia," Aquatic Botany, vol. 62, no. 3, pp. 177–188, 1998.

22. M. A. Khan and I. A. Ungar, "Alleviation of salinity stress and the response to temperature in two seed morphs of Halopyrum mucronatum (Poaceae)," Australian Journal of Botany, vol. 49, no. 6, pp. 777–783, 2001.

23. G. Ne'eman, N. Henig-Sever, and A. Eshel, "Regulation of the germination of Rhus coriaria, a post-fire pioneer, by heat, ash, pH, water potential and ethylene," Physiologia Plantarum, vol. 106, no. 1, pp. 47–52, 1999.

24. Z. Noumi, S. Ouled Dhaou, and M. Chaieb, "Seed germination characteristics of Periploca angustifoliaLabill. and Rhus tripartita (Ucria) Grande: Effects of temperature, salinity and water stress," Acta Botanica Gallica, vol. 157, no. 2, pp. 317–327, 2010.

25. M. A. Chuang, W. H. Zhang, and L. X. Cheng, "Effects of iso-osmotic potential salt and water stress on the seed germination of Periploca sepium," Bulletin of Botanical Research, vol. 28, pp. 465–470, 2003.

26. S. Talbi, A. Ferchichi, M. Debouba, and E. Lefi, "Effect of osmotic stress ($PEG_{6000}$) on final germination percentage and median germination time of Plantago albicans," Revue des Régions Arides, vol. 24, pp. 51–54, 2009.

27. S. Talbi, Effet du stress hydrique sur le comportement physiologique et morphologique de trois espèces pastorales des zones arides tunisiennes: Lotus creticus, Plantago albicans et Rhanterium suaveolens [M.S. thesis], Université of 7 Novembre, Carthage, Tunisia, 2008.

28. M. L. Sharma, "Stimulation of drought and its effect on germination of five pasture species," Agronomy Journal, vol. 65, pp. 962–987, 1973.

29. K. C. Hodgkinson, "The shrubs of poplar box (Eucalyptus populnea) lands and their biology," Australian Rangelands Journal, vol. 1, pp. 280–293, 1979.

30. Y. Gutterman, Survival Strategies of Annual Desert Plants: Adaptation of Desert Organism, Springer, New York, NY, USA, 2002.

31. D. V. Peláez, R. M. Bòo, and O. R. Elía, "Emergence and seedling survival of caldén in the semiarid region of Argentina," Journal of Range Management, vol. 45, pp. 564–568, 1992.

32. R. Zouaoui, Analyse et Description Botanique, Morphologique et Ecophysiologique de quelques Espèces Menacées de Disparition: Ziziphus lotus, Panicum turgidum, Helianthemum lippii sessiliflorum, Arthrophytum schmittianum, Dichantium annulatum et Argyrolobium uniflorum du Parc National de Bouhedma [M.S. thesis], Université of El Manar, Tunis, Tunisia, 2007.

# CHAPTER 5

# Towards the Improvement of Fruit-Quality Parameters in Citrus under Deficit Irrigation Strategies

**A. Aguado, J. Frías, I. García-Tejero, F. Romero, J. L. Muriel, and N. Capote**

IFAPA Centro Las Torres-Tomejil, Carretera Sevilla-Cazalla Km 12'2, Alcalá del Río, 41200 Sevilla, Spain

## ABSTRACT

Water shortage is becoming a severe problem in arid and semiarid regions worldwide, reducing the availability of agricultural land and water resources. In Spain, citrus is one of the most economically important crops, with 74,000 ha devoted to its cultivation. Since water resources are increasingly more insufficient, the efficient use of water is becoming more essential. Deficit irrigation in many agricultural crops has frequently proved to be an efficient tool for improving water-use efficiency. This paper examines the effects a deficit irrigation during the ripening period on yield and the most representative fruit quality properties. The study was conducted during two consecutive years (2009-2010) in a commercial 12-year-old orange orchard (Citrus sinensis L. Osb. cv. Navelina) grafted onto Carrizo citrange (Citrus sinensis L. Osb. × Poncirus Trifoliata L. Osb.). A regulated deficit irrigation (DI) was applied, which was fully irrigated during the flowering and fruit-growth stage, and during the ripening period it was subjected to a water-stress ratio of 0.75. A control treatment was established, this being irrigated at 100% of crop evapotranspiration. Along the water stress period, it was tested the temporal evolution of the main organoleptic and nutraceutical fruit properties (color index, the total soluble solids, titrable acidity, maturity index, rind weight, juice weight, the ratio of juice weight versus fruit weight, and the total C vitamin and flavonoids). It was not observed a descend in juice content, or fruit weight, or in the final yield. In terms of fruit organoleptic and nutraceutical properties, there were not detected negative effects in the studied properties. Furthermore, DI treatment showed higher values of maturity index than control treatment, which can be considered as a positive aspect in the fruit quality. Considering these results, we can affirm that a moderate water stress applied during the maturity period is a sustainable strategy for saving water, increasing the irrigation productivity and obtaining fruits with similar properties to those without deficit irrigation.

## 1. INTRODUCTION

Citrus is one of the most important crops in Andalusian agriculture, with 74,000 ha devoted to its cultivation. In the last 25 years, this surface has grown by 2.04% annually; and it has therefore been promoted that the citrus production will be close to 2.3 million tons by 2015 (30% of the national production). Of the total area of citrus in Andalusia, 51,500 ha are dedicated to sweet orange, being the Navelina variety the second most representative of them with 13,000 ha García-Tejero [1]. Between the most limiting factors for agriculture in Andalusia are its climatic conditions, with a Mediterranean-dry climate, characterized by the irregularity and scarcity of rainfall, and a high evapotranspiration, closely to 1,600 mm, coinciding the months with the largest evapotranspiration levels (June, July, august, and September) with those in which the rainfall are closely to zero.

Considering that water is a scarce resource in Spain, the irrigated agriculture accounting for 80% of global water consumption [2]; it is important to apply different management strategies for increasing the water-use efficiency. In this context, several contributions have documented the advantages of deficit irrigation strategies to improve the efficiency of water use and fruit quality in citrus trees: [3–9]. Shalhevet and Bielorai [10] and Doorenbos and Kassam [11] established that response of the citrus trees to water stress depends mainly on the phenological stage of the crop. Several authors have pointed out the great importance of the growth periods during which water stress is imposed, as well as the different climatic characteristics of the orchard. In this sense, during fruit-growth stage, a water-stress-induced results in a descend of fruit diameter, an increment in fruit drop, and a minor amount of juice per weight unit, this not being a recommendable strategy [12]. In this sense, García-Tejero et al. [13] reported that since the point of view of the irrigation water-use efficiency, it was more useful the management of limited irrigation water amounts during the ripening period, because in this case a water stress is not reflected in a decrease in the final yield, although it could be reflected in the fruit-quality parameters. In this context, the quality of citrus fruit is affected by water stress due to changes in juice properties, such as increases in sugar concentration and acidity, which could reduce in some cases the fruit quality [14]. Several authors such as Ginestar and Castel [4], González-Altozano and Castel [5] or García-Tejero et al. [15] reported that total soluble solids (TSS) or titratable acidity (TA) levels are increased when a moderate water-stress is applied during the maturity period, and this fact may affect to the ratio of these characters, and hence, the time of the harvest [16]. Navarro et al. [14] argued that a moderate water-stress induced higher carbohydrates concentration in juice than unstressed plants; and this fact was independent of fruit size and juice content. In this line, it was related with the main factor involved in this change. Thus, passive dehydratation of juice sacs was not the primary cause of differences in sugar accumulation as it was stated by Barry et al. [17] or Yakushiji et al. [18] among others. Not only classical parameters such as sugar content or juice acidity are affected by water deficit. Navarro et al. [14] found significant increments in ascorbic acid content, although other authors such as Buendía et al. [19] did not find a direct relationship between deficit irrigation and the content of vitamin C.

Considering the previous researches by Pérez-Pérez et al. [20] and García-Tejero et al. [15], we hypothesised that a water-stress-induced during the ripening period can improve or keep the fruit organoleptic properties, without promoting significant descends in final yield and carrying remarkable water saving. The objective of this work was to determine the main effects of deficit irrigation when this is applied during the ripening period on the fruit quality, yield and irrigation water productivity in a commercial orchard of "Navelina" sweet orange.

## 2. MATERIAL AND METHODS

### 2.1. Experimental Site

The trial was conducted during two consecutive years (2009-2010), in a commercial 12-years-old orchard of mature orange trees (Citrus sinensis, L. Osb. cv. Navelina) grafted onto Carrizo citrange (Citrus Sinensis, L. Osb. × Poncirus trifoliata Raf.), located in El Campillo, (Huelva), SW Spain (37°39′50.15″ N; 6°35′53.25″ W). The trees were planted spaced at 6 × 4.5 m and drip irrigated by two pipe lines with pressure-compensating emitters. The average tree height was 3 m, with a canopy diameter of 3.5 m. The soil is an endoleptic cambisol [21], built on slate rock; with 0.6 m of useful deep, and roots growing predominantly within 0.4 m from surface. The soil is classified as a sandy loam with 0.52, 0.33 and 0.15 g kg$^{-1}$ of sand, silt, and clay, respectively, with a field capacity and permanent wilting point of 0.17±0.02 and 0.10±0.03 m$^3$ m$^{-3}$; and a bulk density of 1.31±0.02 g cm$^{-3}$.

The local climate is Mediterranean dry, with an average potential evapotranspiration (ET0) of 1,371 mm yr$^{-1}$and annual rainfall of 780 mm, distributed mainly from late autumn to early spring. The thermal range was broad, with average temperatures being 10°C in winter and 27°C in summer, often exceeding 40°C during the hours of maximum solar radiation.

## 2.2. Experimental Design

The experimental design was a randomized complete block, with four replicates per treatment. The experimental unit had three rows, with five trees per row, and the three central trees (central row) were established for measuring fruit properties and final yield.

A deficit-irrigation treatment (DI) was applied during the ripening period, with a water stress ratio of 0.75, defined as the ratio between the actual water-limited supplied in this treatment versus the water supplied to control treatment. Also, a fully irrigated treatment at 100% of crop evapotranspiration ($ET_c$) was considered as a control. Thus, in 2009 year the water-restriction period was from 301 day of the year (DOY) to harvest (322 DOY), ranging this period from 308 to 330 DOY in 2010.

The seasonal values of ET0 were determined by the modified Penman equation [22]. Climatic data were recorded at an automated weather station near the orchard, using the Doorenbos and Pruitt [23] equation for estimating the real irrigation rates required by the crop

$$\text{Irrigation (mm)} = \sum_{i=1}^{7} [ET_0 \cdot K_C \cdot K_r] - \text{rain}, \qquad 1)$$

where $KC$ is the crop coefficient, and $Kr$ is the reduction coefficient, calculated as twice the ratio of the shaded surface area at noon. In our case, a $Kr$ of 0.6 and a $KC$ ranging between 0.5 from March to May, 0.55 from June to October and 0.5 in November and December, respectively.

According to this, Table 1 shows the irrigation water applied during the studied period for each treatment.

## 2.3. Plant Measurements and Statistical Analysis

During the studied period three different samples were established in each season (301, 315, and 322 DOY in 2009 and 308, 316, 330 DOY in 2010). For each sample there were determined 11 fruit-quality parameters, analyzing 5 fruits per tree. The external fruit colour index

(CI) was measured using a tristimulus colour difference meter (Minolta CR-300 colorimeter), at three locations around the equatorial plane of the fruit and two in the polar plane. The Hunter parameters $a*$, $b*$ and $L*$ were reported by the colorimeter, obtaining the CI using the following equation:

$$CI = \frac{1000 \cdot a}{L \cdot b}. \quad (2)$$

$L*$ parameter gives a value of the luminance or brightness of the sample. The $a*$ parameter indicates the area of variation between red and green spectrum; $b*$ parameter is referred to the area of variation between yellow and blue spectrum.

**Table 1.** Irrigation water applied (mm) in each treatment during the studied period.

| Period (DOY) | Year | | | |
|---|---|---|---|---|
| | 2009 | | 2010 | |
| | Control | DI | Control | DI |
| 1–31 | 0 | 0 | 0 | 0 |
| 32–59 | 7,7 | 7,7 | 0 | 0 |
| 60–90 | 17,2 | 17,2 | 3,6 | 3,6 |
| 91–120 | 21,4 | 21,4 | 11,4 | 11,4 |
| 121–151 | 44,1 | 44,1 | 44,6 | 44,6 |
| 152–181 | 70,8 | 70,8 | 65 | 65 |
| 182–212 | 82,5 | 82,5 | 98,6 | 98,6 |
| 213–243 | 93,5 | 93,5 | 95 | 95 |
| 244–273 | 81 | 81 | 83 | 83 |
| 274–304* | 72,8 | 70,4 | 56 | 56 |
| 305–334* | 21,4 | 15,6 | 20 | 14,6 |
| 335–365 | 0 | 0 | 0 | 0 |
| Total | 512,4 | 504,2 | 477,2 | 471,8 |

DOY: day of the year; Control: control treatment; DI: deficit irrigation treatment; *periods in which DI was applied.

The content of total soluble solids (TSS) in the juice was measured with a refractometer, and titrable acidity (TA) was determined by titration with NaOH and phenolphthalein indicator in 100 mL of juice. Maturity index (MI) was expressed as the relation TSS/TA. Rind weight (RW), juice weight (JW), and the ratio of juice weight versus fruit weight (JR) were determined using a digital scale with a precision of ±0.01 g. Total C vitamin (C-vit) and total flavonoids (TF) were determined by liquid chromatography (HPLC), expressed as mg of ascorbic acid and mg of hesperidin in 100 mL of juice, respectively.

At harvest, fruit yield was determined for each tree by weighing the orange fruits. One sample of 100 fruits per tree was collected to determine average fruit weight.

An exploratory and descriptive analysis was made of yield and its components, followed by an analysis of variance (ANOVA) with a mean separation analysis applying a Tukey's test. SPSS statistical software package was used (SPSS 15.2 for Windows, 2006, SPSS Inc., Chicago, IL).

# 3. RESULTS AND DISCUSSION

## 3.1. Fruit Yield and Irrigation Water Productivity

There was no clear response of the tree yield related to deficit irrigation (Figure 1). During 2009, there was a yield reduction in DI treatment nearly to 5%, this not being considered statistically significant. Additionally, during 2010, the highest values were recorded in stressed treatment, with an increment closely to 30%. Newly, this difference was not statistically significant. Velez et al. [24] were unable to detect any significant differences in either the final production or the fruit weight or in the number of fruits per tree, in response to a DI strategy of cultivation in "Clementine of Nules," when water stress was applied during the maturity period.

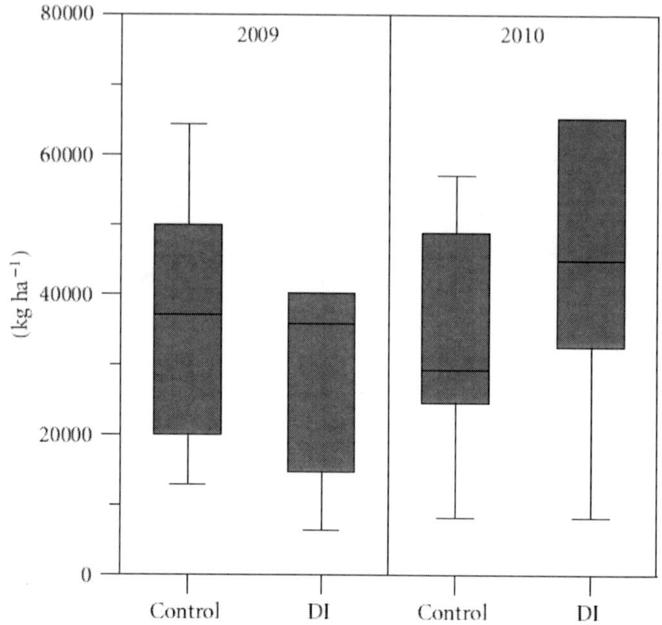


**Figure 1.** Effects of deficit irrigation on yield. Control, control treatment fully irrigated at 100% $ET_c$; DI, deficit irrigation treatment. Vertical lines represent the error bars in each treatment.

In our case, the DI strategy imposed in this experience did not show significant effects on yield, further more, there was not observed an accumulative effect of water stress in subsequent years, and in the second year DI showed a higher yield, with an increment of 11,000 kg $ha^{-1}$. These results are in line with those showed by García-Tejero et al. [13], although these differences cannot be considered significantly different, in economic terms, this improvement in favour of deficit irrigation strategy is important enough to be taken into account by the farmer. These results are in line with those reported by Pérez-Pérez et al. [20] and Castel and Buj [25]. The first of them did not evidence significant effects of deficit irrigation in "lane late" sweet orange when water stress was imposed during the third phase of fruit growth, whereas the second did not find significant yield reduction when water stress was applied in the ripening period in sweet orange cv. "Salustiana".

The lack of effects of DI on final yield along with the water savings promoted increments on irrigation water productivity, this being especially remarkable during 2010 year (Figure 2). The fact was due to the major yields values in DI during this year. Normally, when water is withheld, it causes a yield drop and increases in water productivity. When water stress is considered severe, production drastically declines, decreasing water productivity to some extent. An amount of water applied at different phenological periods can increase water productivity, based on the distribution throughout the irrigation periods [15].

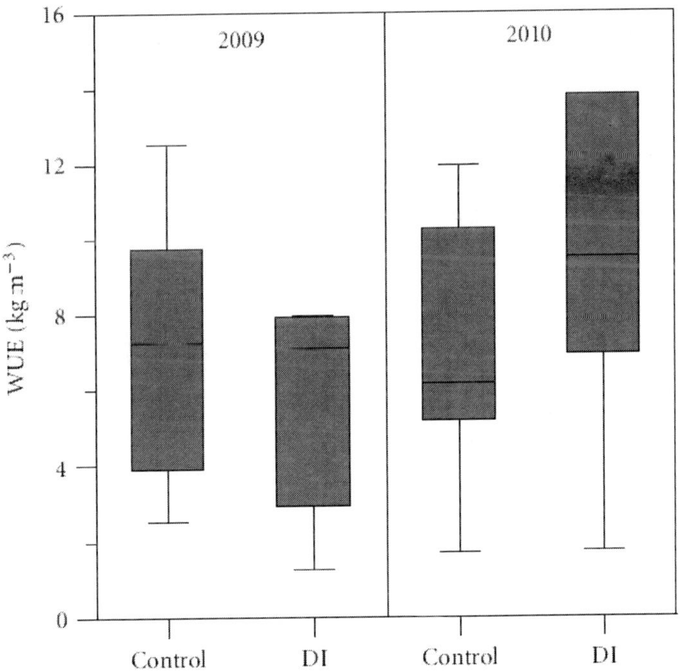


**Figure 2.** Irrigation water productivity in each treatment during the studied period. Control, control treatment fully irrigated at 100% $ET_c$; DI, deficit irrigation treatment. Vertical lines represent the error bars in each treatment.

Increasing crop-water productivity may be a means of achieving efficient and effective water use. In agriculture, the interest is to produce more with less water, because available water for irrigated land is a limiting factor in many world areas [26]. Strategies such as DI have shown that water productivity (WP) can be enhanced [27, 28] and could be associated with acceptable commercial production. Nowadays, the low importance given to improving WP must be related to the reduced water costs in Mediterranean agricultural areas [29], where water represents only less than 10% of the total production costs, a clear contradiction with the Common Agricultural Policy and the water Framework Directive [30].

## 3.2. Impact on Fruit-Quality Parameters

Figure 3 shows the main effects of deficit irrigation in fruit size components. The obtained results evidence non significant effects of water stress in considered parameters. In this context, during 2009 DI treatment did not reflect descends in fruit size and juice weight (Figures 3(a) and 3(b)). Furthermore, this treatment gave better results in these parameters, with increases in fruit and juice weight of 17 and 21%, respectively, with respect to control, although higher values in the ratio between juice and fruit weight in DI with respect to control were not observed (Figure 3(c)). These results evidence larger fruits in DI than control, which is consistent with the observed results of rind weight for each treatment (Figure 3(d)).

During 2010, the obtained results are in line with those in the previous year. There were non significant differences in any parameter, although, again, there were observed increases in juice weight in DI (Figure 3(f)) but fruit weight values between treatments were very similar (Figure 3(e)). This fact promoted larger values of the ratio between juice and fruit weight in DI (a 15% higher), which is particularly remarkable in fruits intended for juice industrial extraction (Figure 3(g)).

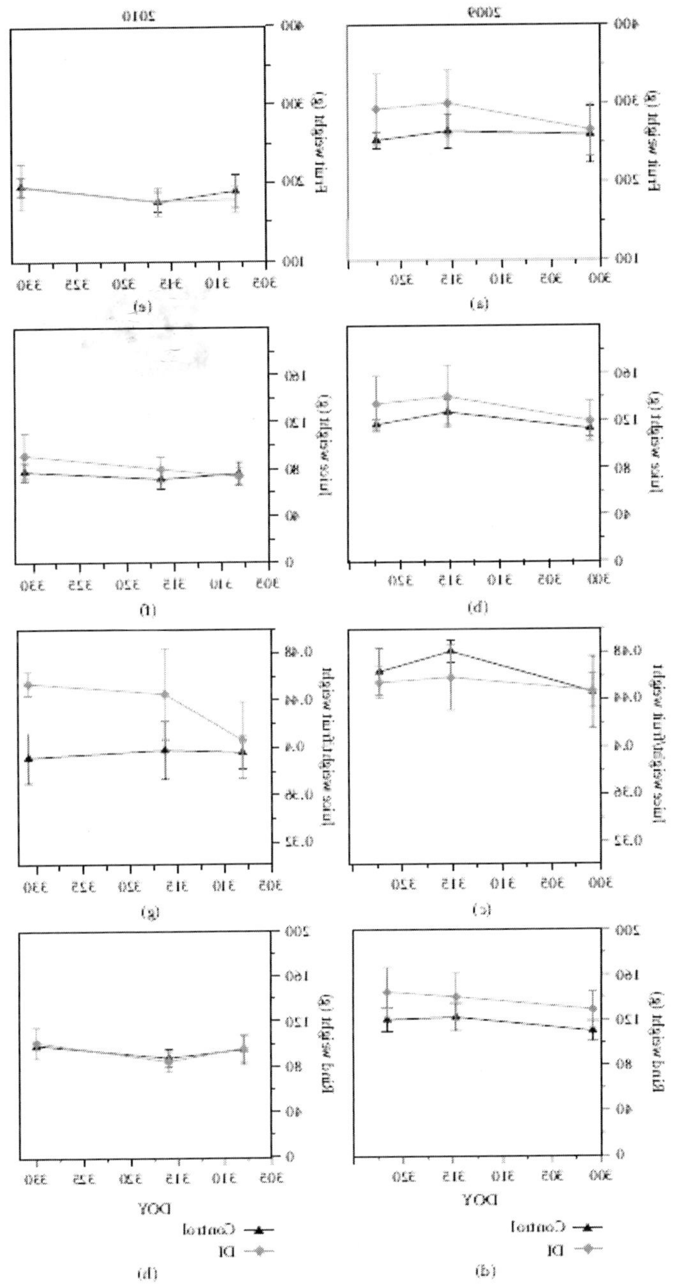

**Figure 3.** Timing evolution of fruit size components. Control, control treatment; DI, deficit irrigation. Vertical lines represent the error bars in each treatment.

On overall, these results argued that a moderate water-stress during the ripening period does not promote negative effects on fruit size and juice content, this stage being a good time for introduce water savings and maintain an optimum fruit development. Ballester et al. [31] did not find significant effects on fruit weight when a moderate water stress was applied during three different stages in "Clementina de Nules." However, other authors such as González-Altozano and Castel [32] in this same cultivar argued that a water stress at the end of summer–autumn could reduce the fruit weight, decreasing the commercial value of them.

Figure 4 shows the main effects of deficit irrigation in fruit quality parameters. The obtained results evidence no significant effects of water stress in considered parameters. In this context, during 2009 DI treatment did not reflected increments in TSS (Figure 4(a)) and TA (Figure 4(b)), fact that also could be observed during the 2010 year (Figures 4(d) and 4(e)). However the ratio between TSS and TA (expressed as MI) was higher in DI treatment during 2009 year (Figure 4(c)). Although this increment was not statistically significant, an increase of MI is considered as a positive point for the commercial value of juice, and advantage an early harvest in sweet orange varieties such as Navelina.

Many studies have been pointed out that water stress in citrus crops affects these properties [15, 33–35]. Verreynne et al. [36] reported that a conventional deficit irrigation strategy with water savings of 60–66% in relation to a control treatment, increased total soluble solids and titrable acidity in "Marisol" Clementines without affecting the juice content or reducing the fruit size. In this line, García-Tejero et al. [13] reported that a severe water stress during the fruit growth and maturity period promoted significant increase in total soluble solids and titrable acidity in different DI strategies such as sustained deficit irrigation and low-frequency deficit irrigation, but these increments are followed by a decrease in maturity index. Moreover, this decline cannot be considered as a positive effect, due to the preservation of equilibrium between the levels of sugars and juice acidity being essential to maintain the juice quality destined for industry. Similar results were reported by Treeby et al. [37] in navel orange, where deficit-irrigation treatments raised total soluble solids and titrable acidity values. In addition, Romero et al. [38] found that severe deficit irrigation raise titrable acidity more than total soluble solids, affecting the final maturity index. The opposite effect was observed in our results and may be related to the water-stress level, duration, and strategy. In this context, Pérez-Pérez et al. [20], studying "lane late" sweet orange found that when water stress was applied during the phase III of fruit growth, there was an increase of total soluble solids and titrable acidity values, although maturity index was not affected. In relation to these results, the total soluble solids content is one of the parameters most affected by a severe water stress. The main cause is not clear, although some authors have explained that this fact can be promoted by a passive dehydratation of juice sacs, accompanied by lower juice content [20].

Finally, the effects of DI during the ripening period in the temporal evolution of external colour index were studied. The obtained results did not show significant differences between treatments (Figure 5). According to Rodrigo et al. [39], during ripening period peel colour changes fast, so that a massive accumulation of carotenoids occurs, from the transformation of chloroplasts to chromoplasts. Navarro et al. [14] argued that a water stress during the fruit growth stage could promote a delay in maturation process. In this same line, if water stress was applied during ripening period, this process occurs in a similar way to full-irrigated treatment.

## 3.3. DI and Nutraceutical Properties

In recent years research activities in agriculture have been aimed to find new production strategies for improving the foods nutraceutical properties. Among the possibilities for improving the quality of them are those designed to achieve an increase in the components recognized as healthy. In this sense, Flavonoids and vitamin C are highly beneficial antioxidant compounds found in many fruits and vegetables, and it's now well established that a plentiful intake of antioxidants through foods, drinks and supplements is vital for optimal human health. It has been recognized that orange juice is an important source of vitamin C and hesperidin for human nutrition. In this sense, we have tried to address a study on the effects of deficit irrigation in the synthesis of these compounds. In line with the previous results, there were no detected significant differences between treatments during the monitoring period. In this sense, moderate deficit irrigation during the ripening period was not reflected with a decrease in the ascorbic acid (Figure 6(a)) and herperidin contents (Figure6(b)).

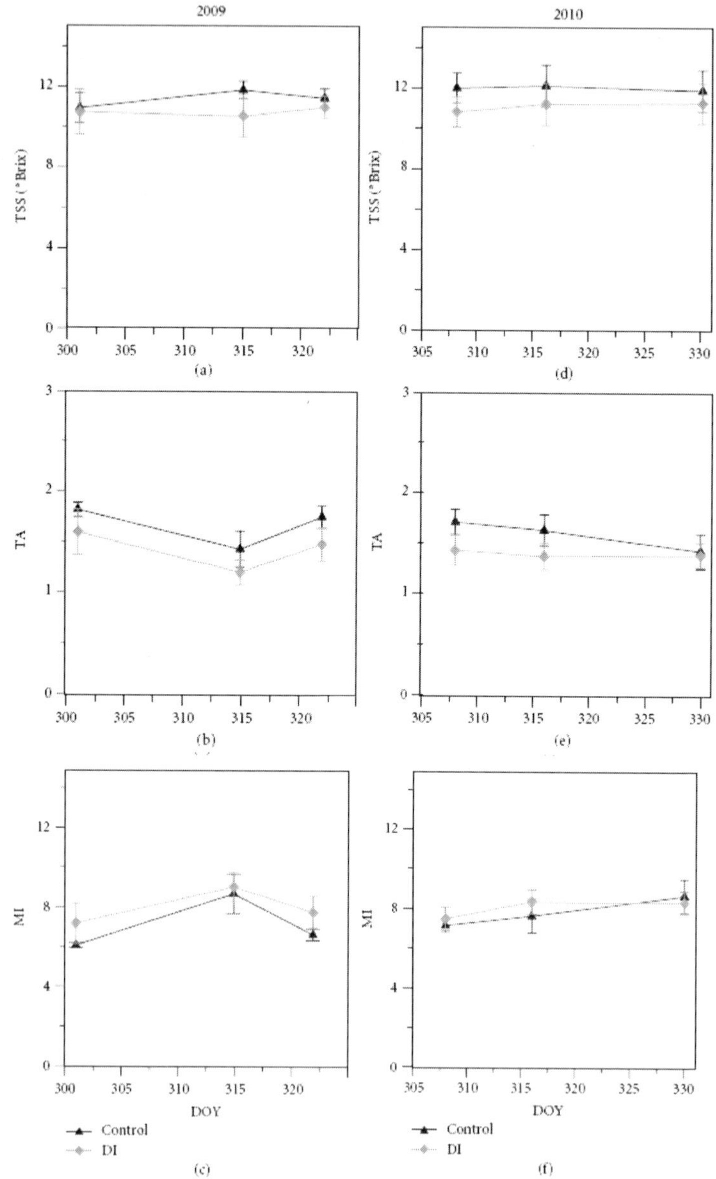


**Figure 4.** Timing evolution of fruit quality parameters. Control, control treatment; DI, deficit irrigation; TSS, total soluble solids; TA, titrable acidity; MI, maturity index. Vertical lines represent the error bars in each treatment.

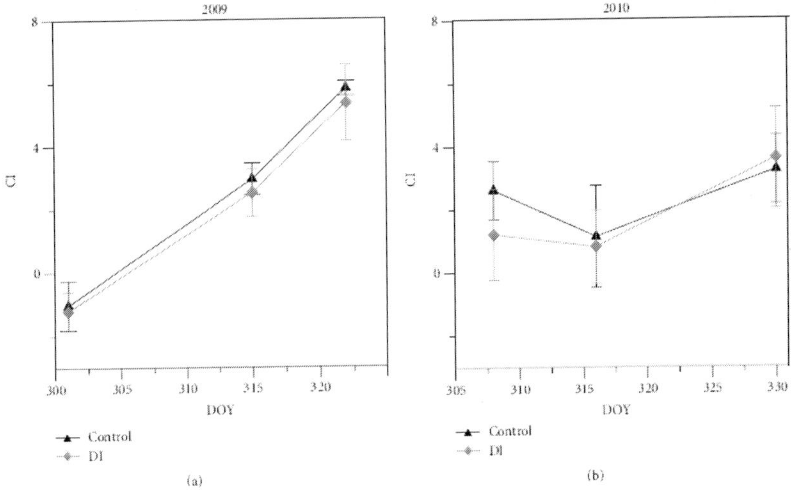


**Figure 5.** Effects of deficit irrigation on colour index (CI) during 2009 (a) and 2010 year (b). Vertical lines represent the error bars in each treatment.

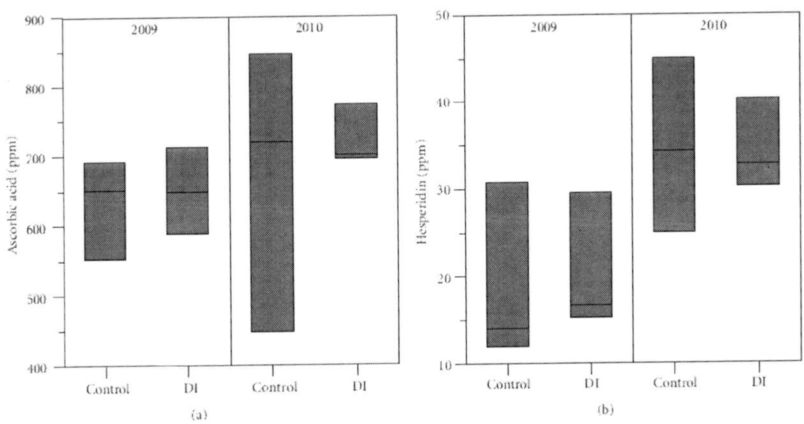


**Figure 6.** Effects of deficit irrigation on ascorbic acid (a) and hesperidin contents (b). Control, control treatment; DI, deficit irrigation treatment. Vertical lines represent the error bars in each treatment.

Navarro et al. [14] showed that DI during stage II of fruit growth promoted significant increments of ascorbic acid with respect to the control juice. They argued that these values could be due to a more pronounced synthesis of vitamin C in response to water stress and/or to the lower fruit water content. If this were true, decreases in juice content would have been detected. These authors did not find significant differences in the contents of ascorbic acid when water stress was applied during phase III of fruit growth, also reflecting higher juice contents in this treatment compared with control treatment. Comparing these results with those obtained in this work, the absence of significant differences would be related with the higher juice contents detected in DI treatment. As similar effects were detected in hesperidin content, this behavior may be due to the same reason that the contents of ascorbic acid.

# 4. CONCLUSIONS

According to the results of the present study, we have not been able to prove that a moderate water stress during the ripening period in sweet orange citrus trees induce a deterioration of fruit quality properties. In this sense, a descend in juice content and in fruit weight was not observed. This fact supports the obtained results on final yield, which was not affected by deficit irrigation, even with an increase on yield values of deficit irrigation.

In terms of fruit organoleptic and nutraceutical properties, there were not detected pernicious effects on them. Moreover, DI treatment showed higher values of maturity index due to the keeping of ratio between the contents of total soluble solids and titrable acidity.

Considering these results, we can affirm that a moderate water stress applied during the last fruit-growth period is a sustainable strategy for saving water, increasing the irrigation productivity and obtaining fruits with similar properties to those without deficit irrigation.

## ACKNOWLEDGMENTS

The authors thank Antonio Caballero for his excellent technical support as personal of Rio Tinto Fruit S.L., where experimental assay has been development. This work was supported by funds from the Project between IFAPA and the Companies Arvensis Agro S.A (CAICEM 09-46), Cultifort S.L. (CC09-009), and Sustineable Agro Solutions S.A. (CEM 09-009).

## REFERENCES

1. I. García-Tejero, Deficit irrigation for sustainable citrus cultivation in Guadalquivir river basin [Ph.D. thesis], Universidad de Sevilla, 2010.

2. MAPA, "Plan Nacional de Regadíos," Horizonte 2008. MAPA. Gobierno de España, 2008.

3. A. Southwick and T. L. Davenport, "Characterization of water stress and low temperature effects on flower induction in citrus," Plant Physiology, vol. 81, pp. 26–29, 1986.

4. C. Ginestar and J. R. Castel, "Responses of young clementine citrus trees to water stress during different phenological periods," Journal of Horticultural Science and Biotechnology, vol. 71, no. 4, pp. 551–559, 1996.

5. P. González-Altozano and J. R. Castel, "Effects of regulated deficit irrigation on "Clementina de Nules" citrus trees. I. Yield and fruit quality effect," Journal of the American Society for Horticultural Science, vol. 74, no. 6, pp. 706–713, 1999.

6. P. González-Altozano and J. R. Castel, "Effects of regulated deficit irrigation on "Clementina de Nules" citrus trees. II. Vegetative growth," Journal of the American Society for Horticultural Science, vol. 75, no. 4, pp. 388–392, 2000.

7. J. L. Muriel, J. A. Jiménez, I. García-Tejero, and I. Vaquero, "Relaciones hídricas en una plantación de naranjos (Citrus Sinensis, L. Cv Navelino) bajo estrategias de riego deficitario mantenido," in VIII Simposium Hispano Portugués de Relaciones Hídricas en las Plantas, D. Morales-Méndez and M. S. Jiménez-Parrondo, Eds., pp. 139–142, Tenerife, Spain, 2006.

8. I. García-Tejero, J. A. Jiménez, J. L. Muriel, and G. Martínez, "Planificación y desarrollo de estrategias de riego deficitario en una plantación de naranjos: influencia de la variabilidad espacial de las propiedades del suelo," in Book of Abstracts, Congreso Nacional de Riegos, Pamplona, Spain, 2007.

9. I. García-Tejero, J. A. Jiménez, M. C. Reyes, A. Carmona, R. Pérez, and J. L. Muriel, "Aplicación de caudales limitados de agua en plantaciones de cítricos del valle del Guadalquivir," Fruticultura Profesional, vol. 173, pp. 5–17, 2008.

10. J. Doorenbos and A. Kassam, "Yield response to water," FAO Irrigation and drainage. Paper no. 33. FAO Rome, Italy, 1979.

11. J. Shalhevelt and H. Bielorai, "Crop water requirements in relation to climate and soil," Soil Science, vol. 125, pp. 240–247, 1978.

12. I. García-Tejero, V. H. Durán-Zuazo, J. Arriaga, and J. L. Muriel-Fernández, "Impact of deficit irrigation on citrus yield," Agronomy for Sustainable Development, vol. 32, no. 3, pp. 651–659, 2012.

13. I. García Tejero, V. H. Durán, J. A. Jiménez, and J. L. Muriel, "Improved water-use efficiency by deficit-irrigation programmes: implications for saving water in citrus orchards," Scientia Horticulturae, vol. 128, no. 3, pp. 274–282, 2011.

14. J. M. Navarro, J. G. Pérez-Pérez, P. Romero, and P. Botía, "Analysis of the changes in quality in mandarin fruit, produced by deficit irrigation treatments," Food Chemistry, vol. 119, no. 4, pp. 1591–1596, 2010.

15. I. García-Tejero, R. Romero, J. A. Jiménez, G. Martínez, V. H. Durán, and J. L. Muriel, "Response of citrus trees to deficit irrigation

during different phenological periods in relation to yield, fruit quality, and water productivity," Agricultural Water Management, vol. 97, no. 5, pp. 614–622, 2010.

16. P. L. Harding, J. R. Winston, and D. F. Fisher, Seasonal Changes in Florida Grapefruit, vol. 886, USDA Technical Bulletin, Washington, DC, USA, 1940.

17. G. H. Barry, W. S. Castle, and F. Davies, "Rootstocks and plant water relations affect sugar accumulation of citrus fruit via osmotic adjustment," Journal of the American Society for Horticultural Science, vol. 129, no. 6, pp. 881–889, 2004.

18. H. Yakushiji, H. Nonami, T. Fukuyama, S. Ono, N. Takagi, and Y. Hashimoto, "Sugar accumulation enhanced by osmoregulation in Satsuma mandarin fruit," Journal of the American Society for Horticultural Science, vol. 121, no. 3, pp. 466–472, 1996.

19. A. Buendía, A. Allende, E. Nicolás, J. J. Alarcón, and M. I. Gil, "Effect of regulated deficit irrigation and crop load on the antioxidant compounds of peaches," Journal of Agricultural and Food Chemistry, vol. 56, no. 10, pp. 3601–3608, 2008.

20. J. G. Pérez-Pérez, J. M. Robles, and P. Botía, "Influence of deficit irrigation in phase III of fruit growth on fruit quality in 'lane late' sweet orange," Agricultural Water Management, vol. 96, no. 6, pp. 969–974, 2009.

21. FAO, World Reference Base for Soil Resources, Rome, Italy, Food and Agriculture Organization of the United Nations, 1998.

22. R. G. Allen, L. S. Pereira, D. Raes, and M. Smith, "Crop evapotranspiration (Guidelines for computing crop water requirements)," FAO Irrigation and Drainage Paper no. 56, FAO, Rome, Italy, 1998.

23. J. Doorenbos and W. O. Pruitt, "Crop water requirements," FAO Irrigation and Drainage paper no. 24, FAO Rome, Italy, 1977.

24. J. E. Velez, D. S. Intrigliolo, and J. R. Castel, "Scheduling deficit irrigation of citrus trees with maximum daily trunk shrinkage," Agricultural Water Management, vol. 90, no. 3, pp. 197–204, 2007.

25. J. R. Castel and A. Buj, "Response of salustiana oranges to high frequency deficit irrigation," Irrigation Science, vol. 11, no. 2, pp. 121–127, 1990.

26. M. H. Ali and M. S. U. Talukder, "Increasing water productivity in crop production—a synthesis,"Agricultural Water Management, vol. 95, no. 11, pp. 1201–1213, 2008.

27. M. H. Ali, M. R. Hoque, A. A. Hassan, and M. A. Khair, "Effects of deficit irrigation on yield, water productivity, and economic returns of wheat," Agricultural Water Management, vol. 92, no. 3, pp. 151–161, 2007.

28. S. K. Jalota, A. Sood, G. B. S. Chahal, and B. U. Choudhury, "Crop water productivity of cotton-wheat system as influenced by deficit irrigation, soil texture and precipitation," Agricultural Water Management, vol. 84, no. 1-2, pp. 137–146, 2006.

29. I. J. Lorite, L. Mateos, and E. Fereres, "Evaluating irrigation performance in a Mediterranean environment: I. Model and general assessment of an irrigation scheme," Irrigation Science, vol. 23, no. 2, pp. 77–84, 2004.

30. M. García-Vila, I. J. Lorite, M. A. Soriano, and E. Fereres, "Management trends and responses to water scarcity in an irrigation scheme of Southern Spain," Agricultural Water Management, vol. 95, no. 4, pp. 458–468, 2008.

31. C. Ballester, J. Castel, D. S. Intrigliolo, and J. R. Castel, "Response of clementina de nules citrus trees to summer deficit irrigation. Yield components and fruit composition," Agricultural Water Management, vol. 98, no. 6, pp. 1027–1032, 2011.

32. P. González-Altozano and J. R. Castel, "Riego deficitario controlado en 'Clementina de Nules'. I. Efectos sobre la producción y la calidad

de la fruta," Spanish Journal of Agricultural Research, vol. 1, pp. 81–92, 2003.

33. H. Bielorai, "The effect of partial wetting of the root zone on yield and water use efficiency in a drip-and sprinkler-irrigated mature grapefruit grove," Irrigation Science, vol. 3, no. 2, pp. 89–100, 1982.

34. I. García-Tejero, J. A. Jiménez, G. Martínez, R. Romero, V. H. Durán, and J. L. Muriel, "Positive impact of regulated deficit irrigation on yield and fruit quality in a commercial citrus orchard (Citrus sinensis (L.) Osbeck, cv. Salustiano)," Agricultural Water Management, vol. 97, no. 5, pp. 689–699, 2010.

35. B. R. Hockema and E. Etxeberria, "Metabolic contributors to drought-enhanced accumulation of sugars and acids in oranges," Journal of the American Society for Horticultural Science, vol. 126, no. 5, pp. 599–605, 2001.

36. J. S. Verreynne, F. Rabe, and K. I. Theron, "The effect of combined deficit irrigation and summer trunk girdling on the internal fruit quality of 'Marisol' Clementines," Scientia Horticulturae, vol. 91, no. 1-2, pp. 25–37, 2001.

37. M. T. Treeby, R. E. Henriod, K. B. Bevington, D. J. Milne, and R. Storey, "Irrigation management and rootstock effects on navel orange [Citrus sinensis (L.) Osbeck] fruit quality," Agricultural Water Management, vol. 91, no. 1–3, pp. 24–32, 2007.

38. P. Romero, J. M. Navarro, P. J. G. Pérez et al., "Deficit irrigation and rootstock: their effects on water relations, vegetative development, yield, fruit quality and mineral nutrition of Clemenules mandarin,"Tree Physiology, vol. 26, no. 12, pp. 1537–1548, 2006.

39. M. J. Rodrigo, J. F. Marcos, and L. Zacarías, "Biochemical and molecular analysis of carotenoid biosynthesis in flavedo of orange (Citrus sinensis L.) during fruit development and maturation," Journal of Agricultural and Food Chemistry, vol. 52, no. 22, pp. 6724–6731, 2004.

# CHAPTER 6

# Performance of 'Valencia' orange (*Citrus sinensis* [L.] Osbeck) on 17 rootstocks in a trial severely affected by huanglongbing

**Kim D. Bowman' Greg McCollum, Ute Albrecht**

US Horticultural Research Laboratory, USDA, ARS,
2001 S. Rock Road, Ft. Pierce, FL 34945, USAa

## HIGHLIGHTS

- In a sweet orange field trial, large differences were noted between rootstocks.
- Highest yields in the trial were on US-942 and US-1516 hybrid rootstocks.
- It was estimated that huanglongbing disease resulted in a 33% yield reduction.

## ABSTRACT

'Valencia' orange (*Citrus sinensis* [L.] Osbeck) was grown on 17 rootstocks through seven years of age and the first four harvest seasons in a central Florida field trial severely affected by huanglongbing (HLB) disease. All trees in the trial had HLB symptoms and were shown by PCR to be infected with *Candidatus* Liberibacter asiaticus (Las). Large differences were noted between rootstocks for many metrics examined, including yield, fruit quality, and tree size. Highest yields in the trial were on US-942 rootstock, which was significantly more productive than trees on the common commercial rootstocks Carrizo, Kuharske, Cleopatra, and Kinkoji. Other new hybrid rootstocks also performed well in this trial strongly affected by HLB, including the rootstock US-1516, which had the second highest cumulative yield, best tree health rating, and lowest number of trees lost due to HLB damage. Comparison of tree performance in this trial with a similar trial conducted prior to the HLB epidemic, allows us to estimate that the disease resulted in a 33% reduction in yield and 21% reduction in tree growth through seven years of age. Use of a tolerant rootstock is suggested as an effective means of ameliorating crop losses to HLB.



## KEYWORDS

Sweet orange; Citrus greening; Fruit breeding


## 1. INTRODUCTION

Profitability of citrus production is strongly influenced by the rootstock used in the planting. Carefully conducted field trials to compare citrus rootstock performance have demonstrated that there can be more than a six-fold difference in yield between the best and worst rootstocks (Wutscher and Bowman, 1999), with rootstocks also exhibiting clear significant effects on tree survival, fruit quality, tree size, and tolerance to numerous biotic and abiotic threats (Barry et al., 2004, Bowman et al., 2003, Castle et al., 2011 and Wutscher and Hill, 1995). Previous reports of rootstock field trials affected by huanglongbing (Albrecht et al., 2012, Nariani, 1981, Cheema et al., 1982 and Van Vuuren and Moll, 1985) have yielded confusing and sometimes contradictory results, in part, because of erratic disease spread and irregular development of symptoms. We describe details of performance for 17 rootstocks in a field trial with 'Valencia' sweet orange established in a commercial grove in central Florida, which became 100% infected with*Candidatus* Liberibacter asiaticus (Las), the believed causal agent for huanglongbing (HLB) in Florida, by the time the planting reached 7 years of age. The results suggest significant differences between rootstocks in their influence on tree tolerance to HLB.

# 2. MATERIALS AND METHODS

## 2.1. Tree propagation

In this trial, seven unreleased U.S. Department of Agriculture hybrid rootstocks were compared with ten commercially available rootstocks (Table 1). Seven of the commercially available rootstocks used (Carrizo, Swingle, US-802, US-812, US-852, US-897, and US-942) were developed and released previously by the USDA breeding program. Seeds of all the selections were grown in Brite Leaf citrus nursery (Lake Panasoffkee, Florida), budded with the scion 'Valencia' clone 55-1 in 2007, and prepared for field planting. The 'Valencia' clone used was obtained from Florida Department of Agriculture and Consumer Services, where it was tested and found free of citrus tristeza virus (CTV), Las, and other known pathogens using conventional indexing methods (Kesinger, 2007).

**Table 1.** Rootstocks tested with parentage or species.

| Rootstock | Parentage/species |
|---|---|
| Carrizo | *Citrus sinensis* (L.) Osbeck × *Poncirus trifoliata* (L.) Raf. |
| Cleopatra | *C. reticulata* Blanco |
| Kinkoji | *C. obovoidea* Takahashi |
| Kuharske | *C. sinensis* × *P. trifoliata* |
| Swingle | *C. paradisi* Macf. × *P. trifoliata* |
| US-801 | *C. reticulata* 'Changsha' × *P. trifoliata* 'English Small' |
| US-802 | *C. grandis* (L.) Osbeck 'Siamese' × *P. trifoliata* 'Gotha Road' |
| US-809 | *C. reticulata* 'Changsha' × *P. trifoliata* 'English Large' |
| US-812 | *C. reticulata* 'Sunki' × *P. trifoliata* 'Benecke' |
| US-827 | *C. reticulata* hybrid 'Rangpur' × *P. trifoliata* |
| US-852 | *C. reticulata* 'Changsha' × *P. trifoliata* 'English Large' |
| US-896 | *C. reticulata* 'Cleopatra' × *P. trifoliata* 'Rubidoux' |
| US-897 | *C. reticulata* 'Cleopatra' × *P. trifoliata* 'Flying Dragon' |
| US-942 | *C. reticulata* 'Sunki' × *P. trifoliata* 'Flying Dragon' |
| US-1503 | *C. grandis* 'African' × *P. trifoliata* 'Flying Dragon' |
| US-1516 | *C. grandis* 'African' × *P. trifoliata* 'Flying Dragon' |
| US-1524 | *C. grandis* 'African' × *P. trifoliata* 'Flying Dragon' |

## 2.2. Field planting

The budded trees were transplanted into the trial in June 2008 at a commercial field site in Polk County, Florida, owned by Wheeler Citrus, at a spacing of 4.4 m × 7.6 m. The experiment included 21 'Valencia' trees on each of the 17 rootstocks, planted in a randomized complete block design, in ten adjacent rows about 200 m long. Border trees with the same scion were planted on each end of the rows and in the two adjacent rows. Soil was Candler sand, with good natural drainage, and a gentle slope. Irrigation in the block was by under-tree microjets.

## 2.3. Field plot management

The rootstock trial planting was adjacent to a commercial block of trees with 'Valencia' scion, and management care of the trial block was the same as the contiguous commercial block. General care of the block was as follows. During the bearing years, fertilizer was applied by broadcast spreader in three equal applications of 12-2-16 with Mg and Zn, at a rate of 72 kg N per acre per year. Weeds were controlled by 3 applications/year of glyphosate (Roundup; Monsanto, St. Louis MO) and Diuron [3-(3,4-dichlorophenyl)-1,1-dimethylurea; Loveland Products, Loveland, CO] in a 2.4 m band within the row, from 2008 through 2012. From 2012 to 2015, weed control was by application of glyphosate within the row as needed. Row middles were developed with bahia grass cover and were managed with periodic mowing. Chemical pest control was applied five times per year using a speed sprayer. The first application was in the spring at 2/3 petal fall and contained copper hydroxide and spirotetramat (Movento; Bayer CropScience, Research Triangle Park, NC). A second treatment, containing oil, copper hydroxide, and abamectin (Agri-Mek; Syngenta Crop Protection, Greensboro, NC), was applied when the May flush reached full expansion. A mixture of thiamethoxam and abamectin (Agri-Flex; Syngenta Crop Protection) was applied three times per year to combat insects and mites.

## 2.4. Tree yield and fruit quality

Yields for groups of three trees (seven replicates of three trees for each rootstock) were measured during harvest in April of each year, 2012–2015, by a cooperative effort of the commercial harvesting crew and USDA research staff. Yield/tree and cumulative yield/tree was calculated based on the trees that remained (surviving trees) in the planting at the time of each season's harvest, except where it is indicated that numbers are based on the number of planted trees (pt). Samples of 24-fruit for quality analysis were collected from each replicate of three trees just before harvest each year. Fruit samples were juiced using an FMC Multifruit Fresh N'squeeze juicer, model POS1 (JBT FoodTech Citrus Systems, Lakeland, FL), and juice quality was analyzed using standard laboratory methods.

## 2.5. Tree size

Tree size measurements of scion trunk cross sectional area (TCA) 5 cm above the graft union, canopy height, and canopy spread were taken in July 2015 using tape measure, height pole, and caliper. Canopy volume was calculated by the formula (diameter$^2$ × height)/4 (Wutscher and Hill, 1995). Yield efficiency was calculated as the ratio of fruit yield per surviving tree in 2015 to canopy volume in July 2015.

## 2.6. Detection of Las and CTV

For PCR detection of Las in trees, eight leaves were collected from each plant in July 2015, two from each cardinal direction using only mature leaves from the latest flush at a height of 1.5–2.0 m above the ground. Petioles and parts of the midrib were severed from leaves and ground in liquid nitrogen with a mortar and pestle. One-hundred mg of ground tissue was used for DNA extraction. DNA was extracted using the Plant DNeasy® Mini Kit (Qiagen, Valencia, CA) according to the manufacturer's instructions, but omitting RNA removal to allow for detection of citrus tristeza virus (CTV) and Las in the same sample. For detection of Las, real-time PCR assays were performed using primers HLBas and HLBr and probe HLBp developed by Li et al. (2006). For normalization, all samples were assayed using primers COXf and COXr and probe COXp (Li et al., 2006). Amplifications were performed over 40 cycles using an ABI 7500 real-time PCR system (Applied Biosystems, Foster City, CA) and the QuantiTect Probe PCR Kit (Qiagen) according to the manufacturer's instructions. All reactions were carried out in a 20 µl reaction volume using 5 µl of extracted DNA/RNA. Plants were considered PCR-positive when normalized $Ct_{LAS}$-values were $\leq 33$.

For quantification of Las, $Ct_{LAS}$-values were converted to copy numbers of Las genomes based on a standard curve created in our laboratory: $y = 12.34–0.32x$, where $y$ = log of Las copy number and $x$ = normalized $Ct_{LAS}$-value. Ct-values not determined after 40 cycles were assigned a value of 41. Since three copies of the 16s rDNA gene are present in the Las genome (Duan et al., 2009), copy numbers were divided by three and data were expressed as numbers of Las genomes per milligram of plant tissue.

For detection of CTV, real-time PCR assays were performed using primers UTR1 and UTR2 and probe 181T (Bertolini et al., 2008). Amplifications were carried out in an AB7500 real-time PCR system using the QuantiTect Probe RT-PCR Kit (Qiagen) according to the manufacturer's instructions and using 2 μl of extracted DNA/RNA in a 20 μl reaction volume.

## 2.7. Foliar disease symptoms

Trees were examined for chlorosis, blotchy mottle, green islands, or other foliar abnormalities presumably associated with HLB in April 2015, and scored on a scale of 0–5 (0 = no foliar disease symptoms; 1 = <10% foliar disease symptoms, dense canopy; 2 = 10–25% foliar disease symptoms, dense canopy; 3 = 25–50% foliar disease symptoms, mostly dense canopy/few areas with open canopy/few rabbit ears; 4 = 50–75% foliar disease symptoms, open canopy/rabbit ears abundant; 5 = >75% foliar disease symptoms, canopy very open/rabbit ears very abundant).

## 2.8. Statistical analyses

Data were tested by analysis of variance using Statistica ver 10.0 (StatSoft, Tulsa, Okla.) and comparison of the means was by Duncan multiple range test at $P < 0.05$. Foliar disease symptom ratings were tested by Kruskal–Wallis ANOVA by Ranks.

**Table 2.** Yield of 'Valencia' trees on 17 rootstocks in a Polk county trial.

| Rootstock | Yield 2012 (kg/tree) | Yield 2013 (kg/tree) | Yield 2014 (kg/tree) | Yield 2015 (kg/tree) | Yield 2012–15 (kg/tree) |
|---|---|---|---|---|---|
| US-942 | 43 a | 67 a | 66 a | 52 abc | 227 a |
| US-1516 | 32 bc | 60 abc | 54 abc | 65 a | 211 ab |
| US-896 | 36 ab | 64 ab | 61 ab | 44 abc | 205 abc |
| US-1503 | 29 bcd | 47 cde | 60 ab | 61 ab | 197 abc |
| US-1524 | 27 bcd | 48 cde | 56 abc | 63 ab | 193 abc |
| Swingle | 27 bcd | 47 cde | 50 a–d | 56 abc | 180 a–d |
| US-802 | 27 bcd | 59 bcd | 47 a–d | 43 abc | 176 bcd |
| Carrizo | 25 cd | 49 b–e | 48 a–d | 45 abc | 168 bcd |
| US-852 | 31 bc | 49 b–e | 41 a–d | 45 abc | 167 bcd |
| Kuharske | 32 bc | 45 cde | 47 a–d | 42 a–d | 166 bcd |
| US-809 | 28 bcd | 50 b–e | 40 bcd | 46 abc | 164 bcd |
| US-812 | 35 abc | 47 cde | 35 cd | 44 abc | 162 bcd |
| Cleopatra | 16 e | 34 e | 51 a–d | 57 abc | 158 cd |
| US-827 | 27 bcd | 42 de | 34 cd | 38 bcd | 141 d |
| US-897 | 34 abc | 38 e | 32 cd | 33 cd | 137 d |
| Kinkoji | 19 de | 39 e | 38 bcd | 34 cd | 130 d |
| US-801 | 36 abc | 47 cde | 27 d | 19 d | 128 d |

Mean separations for significant ANOVA within columns were by Duncan's multiple range test at $P < 0.05$.

# 3. RESULTS

## 3.1. Fruit yield

Fruit yield was measured during the harvest seasons 2012–2015, and across all rootstocks averaged 43 kg per tree per season. There were significant differences among the rootstocks in fruit yield during each of the four harvest seasons, 2012–2015, with US-942 having the highest production each of the first three seasons and highest cumulative yield (Table 2), and averaging 57 kg of fruit per tree per season. Other rootstocks in the trial that are commonly used in Florida, including Swingle, US-802, Carrizo, Kuharske, US-812, and Cleopatra, were intermediate in yield per tree, ranging from 39 to 45 kg/tree/season. The two rootstocks with the lowest yield in the trial were Kinkoji and US-801, both averaging 32 kg fruit/tree/season. Among the other rootstocks that are not commercially available, US-1516, US-1503, US-1524, and US-896 yielded particularly well, ranging from 48–53 kg/tree/season.

## 3.2. Fruit quality

Fruit quality parameters were measured during each of the seasons 2012–2015. Analysis of variance for main effects of harvest year and rootstock were highly significant for each of the fruit quality parameters averaged over the four years (Table 3). Although harvest year by rootstock interactions for fruit weight, total acidity and juice color were significant for each parameter, the interaction effect was a minimal component of the variance. Largest mean fruit weight (222 g) was obtained with Kuharske, and smallest (173 g, 180 g and 184 g) with US-801, US-897 and US-896, respectively, while most of the other hybrids produced fruit not significantly different in weight from Kuharske.

**Table 3.** Fruit quality of 'Valencia' trees on 17 rootstocks in a Polk county trial assessed for the years 2012–2015.

| | Fruit wt (g) | Juice (%) | TSS (%) | Acid (%) | Ratio | Juice color (CN) |
|---|---|---|---|---|---|---|
| Rootstock | | | | | | |
| US-896 | 184 d | 57.2 ab | 9.97 a | 0.84 a–c | 12.0 bc | 38.4 ab |
| US-801 | 173 d | 55.6 b–d | 9.97 ab | 0.88 a | 11.5 c | 38.3 ab |
| US-897 | 180 d | 56.5 a–c | 9.81 a–c | 0.87 ab | 11.5 c | 38.2 b |
| Carrizo | 205 bc | 55.7 b–d | 9.74 a–d | 0.80 c–e | 12.3 a–c | 38.4 ab |
| US-942 | 204 bc | 56.9 ab | 9.73 a–d | 0.81 b–d | 12.0 bd | 38.3 ab |
| US-852 | 217 a–c | 56.0 a–d | 9.72 a–d | 0.79 c–e | 12.5 ab | 38.2 b |
| Kuharske | 222 a | 55.9 b–d | 9.69 a–e | 0.74 ef | 13.2 a | 38.6 a |
| Kinkoji | 202 c | 56.7 a–c | 9.64 a–e | 0.78 c–e | 12.4 a-c | 38.2 b |
| US-809 | 202 c | 57.5 a | 9.61 a–e | 0.78 c–e | 12.4 a-c | 38.3 ab |
| US-1503 | 212 a–c | 56.0 a–d | 9.60 a–e | 0.82 b–d | 11.9 bc | 38.3 ab |
| US-1524 | 210 a–c | 55.3 cd | 9.59 a–e | 0.78 c–e | 12.4 a-c | 38.4 ab |
| US-802 | 217 a–c | 54.7 d | 9.55 a–e | 0.81 b–d | 11.9 bc | 38.3 ab |
| US-827 | 215 a–c | 56.2 a–d | 9.50 a–e | 0.84 a–c | 11.6 c | 38.2 b |
| US-1516 | 212 a–c | 54.8 d | 9.41 b–e | 0.76 de | 12.5 ab | 38.6 a |
| US-812 | 210 a–c | 56.2 a–d | 9.35 c–e | 0.77 de | 12.4 a-c | 38.4 ab |
| Swingle | 213 a–c | 56.3 a–d | 9.19 de | 0.74 ef | 12.5 ab | 38.6 a |
| Cleopatra | 219 ab | 56.2 a–d | 9.15 e | 0.71 f | 13.1 a | 38.5 ab |
| Year | | | | | | |
| 2012 | 228 a | 57.4 a | 9.1 d | 0.75 c | 12.3 b | 38.5 a |
| 2013 | 217 bt | 55.3 c | 10.3 a | 0.81 b | 12.8 a | 38.5 a |
| 2014 | 183 d | 56.3 b | 9.6 b | 0.87 a | 11.2 c | - |
| 2015 | 194 c | 55.4 c | 9.4 c | 0.75 c | 12.7 a | 38.1 b |
| Factor (df) | | | | | | |
| Replicate (6) | 1.87$^{***}$ | 0.002$^{*}$ | 2.27$^{**}$ | 0.035$^{***}$ | 3.5 ns | 0.44$^{*}$ |
| Year (3) | 29.1$^{***}$ | 0.011$^{***}$ | 28.15$^{***}$ | 0.406$^{***}$ | 64.2$^{***}$ | 4.34$^{***}$ |
| Rootstock (16) | 3.11$^{***}$ | 0.002$^{***}$ | 1.48$^{**}$ | 0.057$^{***}$ | 6.8$^{***}$ | 0.36$^{***}$ |
| Year × Rootstock (48) | 0.62$^{***}$ | 0.001 ns | 0.8 ns | 0.012$^{*}$ | 2.0 ns | 0.28$^{*}$ |
| Error (400) | 0.35 | 0.001 | 0.8 | 0.009 | 2.1 | 0.18 |
| Total (473) | | | | | | |

Fruit quality data presented for individual rootstocks are averaged over years 2012–2015. Mean separations for significant ANOVA within columns were by Duncan's multiple range test at $P < 0.05$. $^{*}$, $P < 0.05$; $^{**}$ $P < 0.01$; $^{***}$ $P < 0.001$.

The greatest difference in percent juice was between fruit produced on US-809 (57.5%) compared with US-802 (54.7%) and US-1516 (54.8%), which were not significantly different. There was considerable similarity in percent juice among the additional rootstocks. Total soluble solids (TSS) ranged from a low of 9.15% for fruit produced on Cleopatra to a high of 9.97% for fruit produced on US-896. Fruit produced on Swingle had TSS of 9.19% which was not significantly different from Cleopatra. Most of the hybrid rootstocks produced fruit with TSS lower than, but not significantly different from, US-896. Total acidity was lowest (0.71%) in fruit produced on Cleopatra and highest (0.88%) in fruit produced on US-801. Highest TSS:Acid ratios (13.2:1 and 13.1:1) were in fruit produced on Kuharske and Cleopatra, and lowest (11.5:1) on US-897 and US-801. Although main effects of rootstock on juice color were statistically significant, the range in color number was only from 38.2 to 38.6, indicating no practical difference among the rootstocks.

Although all fruit quality parameters differed significantly among the harvest years, clear trends were only seen for fruit weight and TSS. Mean fruit weights decreased from a maximum of 228 g in 2012 to a minimum of 194 g in 2015, representing a 15% decrease over the four years in which data were collected. The lowest TSS (9.1%) were recorded in 2012 (Table 3) and highest (10.3%) in 2013. The low TSS in 2012 may have been a function of tree age, as it is known that citrus fruit TSS is typically lower from less mature than from more mature trees. Between 2013 and 2015 mean TSS dropped from 10.3 to 9.4, a 15% decrease.

## 3.3. Tree size and yield efficiency

Scion trunk cross sectional area (TCA), canopy height, canopy spread, and canopy volume were influenced significantly by rootstock when measured at 7 years of age (Table 4). Trees on US-802 rootstock were the largest by all measures of size except TCA, where trees on Cleopatra rootstock were largest. Trees on US-897 rootstock were the smallest by all measures of size. Trees on US-801, the rootstock that appeared most strongly affected by HLB, were intermediate in size, and the canopy volume of trees on this rootstock were not significantly different from those on Cleopatra, Carrizo, Kuharske, Kinkoji, or Swingle. Yield efficiency values ranged from 3.73 $kg/m^3$ for US-801 to 11.31 $kg/m^3$ for Swingle.

**Table 4.** Tree size of 'Valencia' trees on 17 rootstocks in a Polk county trial measured in 2015.

| Rootstock | ScionTCA ($cm^2$) | Canopy height(m) | Canopy diameter (m) | Canopy volume ($m^3$) | Yield efficiency ($kg/m^3$) |
|---|---|---|---|---|---|
| US-802 | 97.3 a | 2.84 a | 3.23 a | 7.44 a | 5.78 |
| US-1503 | 89.2 ab | 2.71 a–c | 3.05 a–c | 6.47 ab | 9.43 |
| US-812 | 81.0 b–d | 2.67 a–d | 3.06 a–c | 6.41 a–c | 6.86 |
| US-1516 | 86.7 a–c | 2.63 a–e | 3.07 ab | 6.28 a–d | 10.35 |
| US-942 | 80.6 b–d | 2.67 a–d | 3.03 a–d | 6.23 a–e | 8.35 |
| US-827 | 77.4 b–e | 2.75 ab | 3.00 b–d | 6.20 b–e | 6.13 |
| US-1524 | 80.8 b-d | 2.66 a–d | 3.01 b–d | 6.08 b–e | 10.36 |
| Cleopatra | 98.6 a | 2.44 de | 2.96 b–d | 5.53 b–f | 10.31 |
| Carrizo | 79.3 b–d | 2.46 c–e | 2.86 b–f | 5.35 b–f | 8.41 |
| Kuharske | 89.2 ab | 2.56 b–e | 2.87 b–f | 5.35 b–f | 7.85 |
| Kinkoji | 86.7 a–c | 2.49 c–e | 2.86 b–f | 5.13 c–g | 6.63 |
| US-801 | 71.8 c–f | 2.53 b–e | 2.85 b–f | 5.10 d–g | 3.73 |
| US-896 | 64.0 e–g | 2.39 e | 2.90 b–e | 5.09 d–g | 8.64 |
| US-852 | 76.9 b-e | 2.49 c–e | 2.83 c–f | 5.00 d–g | 9.00 |
| Swingle | 65.6 d–g | 2.42 de | 2.81 d–f | 4.95 e–g | 11.31 |
| US-809 | 60.2 fg | 2.40 ef | 2.71 ef | 4.47 fg | 10.29 |
| US-897 | 53.7 g | 2.17 f | 2.68 f | 3.92 g | 8.42 |

Mean separations for significant ANOVA within columns were by Duncan's multiple range test at $P < 0.05$

## 3.4. CTV and Las infection and HLB effects

The average percent of trees that were CTV positive across all rootstocks was 27.4%. The proportion of trees that were positive by rootstock ranged from 5.6–47.1%. Of the 357 trees originally planted in the trial, 52 trees were removed by the grower between 2009 and harvest in 2015 because of poor health (primarily symptoms of HLB). All of the 305 trees that remained in the trial in July 2015 from the original planting tested positive by PCR for Las, and 304 of the trees showed visible leaf symptoms of HLB (Fig. 1). Mean $Ct_{Las}$ values for the remaining trees differed significantly by rootstock (Table 5), with trees on US-801 having the lowest $Ct_{Las}$ value (and calculated highest Las copy number), and US-897 and US-896 having the highest $Ct_{Las}$ (and calculated lowest Las copy number). Although there were significant differences between $Ct_{Las}$ values for rootstocks, trees on all rootstocks had a large amount of the bacteria present in tissues sampled. Tree removal in the trial by the grower/cooperator was based on visual symptoms of HLB not PCR results, but it can be noted that US-801 also had the largest number of trees removed by the cooperator by July 2015. Although statistical comparisons of the visual tree rating for HLB symptoms in April 2015 were not significant ($P > 0.05$), it can be observed that surviving trees on US-801 had the highest mean value for HLB symptoms at that time.

**Figure 1.** Photo from the trial taken just before harvest, showing a tree with mild (left) and severe (right) symptoms of HLB.

**Table 5.** Infection with Las and effects of HLB for 'Valencia' trees on 17 rootstocks in a Polk county trial.

| Rootstock | Mean Ct Las | Las copy number/mg plant tissue | Trees removed for HLB (%) | Surviving tree health rating | Kruskal–Wallis mean rank[a] |
|---|---|---|---|---|---|
| US-896 | 27.8 a | 301 | 10 | 3.3 | 151 |
| US-897 | 27.7 ab | 374 | 19 | 3.1 | 128 |
| Kuharske | 27.2 a–c | 513 | 0 | 3.3 | 144 |
| US-1516 | 27.0 a–c | 428 | 0 | 3.1 | 117 |
| US-942 | 26.9 a–c | 641 | 14 | 3.4 | 151 |
| US-809 | 26.9 a–d | 590 | 19 | 3.4 | 158 |
| Swingle | 26.8 a–d | 515 | 10 | 3.3 | 140 |
| Carrizo | 26.7 b–d | 713 | 19 | 3.1 | 143 |
| US-852 | 26.6 b–d | 728 | 19 | 3.5 | 159 |
| US-1524 | 26.6 cd | 761 | 14 | 3.3 | 157 |
| US-1503 | 26.6 cd | 639 | 5 | 3.2 | 139 |
| Kinkoji | 26.5 cd | 642 | 14 | 3.4 | 161 |
| US-827 | 26.4 cd | 617 | 29 | 3.6 | 189 |
| Cleopatra | 26.4 cd | 748 | 14 | 3.7 | 185 |
| US-812 | 26.3 cd | 765 | 5 | 3.2 | 135 |
| US-802 | 26.2 cd | 798 | 5 | 3.7 | 182 |
| US-801 | 25.8 d | 1,050 | 52 | 3.8 | 192 |

Mean separations for significant ANOVA within columns were by Duncan's multiple range test at $P < 0.05$.

[a] Kruskal–Wallis ANOVA test was applied to tree health rating, $P = 0.3668$.

# 4. DISCUSSION

This study investigated the performance of 'Valencia' sweet orange on 17 rootstocks in a commercial trial conducted in Polk County, Florida (USA), between 2008 and 2015. The trial was strongly affected by HLB, as indicated by nearly 100% of the trees showing symptoms of HLB and testing positive for Las infection by PCR by 7 years of age. As has been noted previously (Bowman and McCollum, 2015), HLB appears to rarely cause field tree death in Florida, but typically causes a decline in tree health that results in a sparse canopy and little marketable fruit. The resulting unhealthy trees are usually abandoned or removed by citrus growers. In this trial, the grower removed about 15% of the originally planted trees by 7 years of age because of poor performance and pronounced visible symptoms of HLB.

Although this field trial was not originally focused on testing rootstocks for tolerance of huanglongbing disease in Florida, the speed and extent to which it became infected by the pathogen Las created a good opportunity to evaluate rootstock effect on tree tolerance to the disease. Trees in the trial also became infected with CTV, with 27.4% of trees testing positive for the virus at seven years of age, but the isolate of CTV found in the area of the trial is known to have no significant effect on growth of trees on the rootstocks included in this trial (Bowman and Garnsey, 2001). Rootstock was seen to have a significant effect on most of the important traits measured in the trial, including fruit production, fruit quality, tree size, and $Ct_{Las}$ of the scion. The results from this trial provided much better definition of rootstock effects on tree performance under pressure from HLB than our previous report on rootstock field trials with HLB (Albrecht et al., 2012). We believe this is primarily because the previous report relied strongly upon tree growth and HLB symptom scores rather than multi-year yield and fruit quality data, as well as because those trials used fewer replications and trees per rootstock that were more subject to effects from uneven disease spread. Natural spread of Las into field plantings can be erratic and non-uniform over time and area, and this can fatally compromise field trials with insufficient statistical replications. Differences in results between that previous report and this study can also be attributed, in part, to the different location in which the present study was conducted. In this trial, as in many other locations in Florida now, rootstock tolerance to HLB has a dominant effect on overall tree performance. However, it is important to recognize that performance of citrus rootstocks in the field is affected by many different abiotic and biotic factors, and that relative rootstock field performance will often vary considerably between trials in different locations.

Estimation of the quantitative effects of HLB on 'Valencia' tree performance can be obtained by comparison of this study to another rootstock trial with the same scion conducted under very similar conditions between 1991 and 1998, before the HLB epidemic in Florida (Bowman and Rouse, 2006 and Wutscher and Bowman, 1999). Although other factors may play a part in the differences in yield, tree loss, and tree size between the two trials, we believe that the predominant factor in these differences is Las infection and the effect of HLB. A comparison of yield, tree loss, and tree size (TCA) for the five rootstocks in common for the two trials (Table 6) shows a substantially reduced performance in the trial affected by HLB. Yield reduction in the HLB affected trial through the first 7 years (first four harvests) ranged by rootstock from 22–49%, with an average reduction of 33%. Reduction in tree size, as measured by TCA at 7 years of age, ranged from 2 to 34%, with an average reduction in tree size of 21%. Although we have not attempted to calculate the economic impact of HLB in this study, it is clear that the 33% yield loss will have a large effect on profitability for the commercial grower.

Studies of rootstock genotype response to Las infection without grafting and under greenhouse conditions have proven valuable for understanding of Las infection and HLB disease development (Albrecht and Bowman, 2011 and Albrecht and Bowman, 2012), but they appear insufficient to make good predictions about ultimate field performance of grafted trees infected with Las. The results of this study indicate the strong potential to identify significant and commercially meaningful differences between rootstocks in field tolerance to HLB by the use of carefully designed and conducted field trials. Among the commercially available rootstocks, US-942 (Bowman, 2010) appeared to have a clear advantage for commercial use under the conditions of this trial, by yielding the most fruit and having good fruit quality. In the comparison with a second trial without HLB (Table 6), US-942 also appeared to be one of the rootstocks least affected by HLB in terms of cropping and tree growth. As with other rootstocks previously identified as exhibiting improved field tolerance to HLB (Bowman and McCollum, 2015b), we believe the reaction of US-942 to Las infection is best described as tolerance, since sweet orange trees on this rootstock still become infected with Las, maintain relatively high titers of the bacteria, and show leaf symptoms.

Among rootstocks included in the trial that are not commercially available, US-801 deserves special note because it exhibited the lowest yields, smallest fruit size, highest fruit acid, highest proportion of trees removed, lowest $Ct_{Las}$ values, highest Las copy number, and the highest symptom rating of surviving trees. The exceptionally poor performance of US-801 rootstock in this trial suggests an exceptional sensitivity to HLB, since its performance was good in other trials before the introduction of Las to Florida (Wutscher and Hill, 1995). Among other rootstocks in the trial that are not commercially used in Florida, US-1516 and US-896 deserve particular attention because of a combination of high yield and good fruit quality. Trees on these two rootstocks differed significantly in size (TCA), with US-1516 making a larger tree than US-896. It can also be noted that trees on US-1516 had the lowest percent of tree loss through seven years (0%), and lowest HLB symptom score (3.1) of any rootstock in the trial. Since the use of rootstocks which maintain higher yields and good fruit quality despite Las infection appears an effective way of ameliorating the effects of HLB disease, US-1516 was released by USDA in October 2015 (Bowman and McCollum, 2015a).

## ACKNOWLEDGEMENTS

We thank the cooperator, David Wheeler, for his willingness to establish and care for the trial, and for his assistance in collecting data. We thank Emily Domagtoy, Diane Helseth, Sailindra Patel, Kerry Worton, James Salvatore, and Wayne Brown for technical assistance. This research was supported in part by grants from the Florida Citrus Research and Development Foundation. Mention of a trademark, warranty, proprietary product, or vendor does not imply an approval to the exclusion of other products or vendors that also may be suitable.

**Table 6.** Comparison of performance of 'Valencia' on five rootstocks in this trial, with a similar trial conducted in Polk county before the HLB epidemic (Wutscher and Bowman, 1999).

| | Results from trial not affected by HLB[a] | | Results from this trial, affected by HLB | | Comparison of the two trials | |
|---|---|---|---|---|---|---|
| Rootstock | Kg yield per tree | Trees removed (%) | Kg yield per tree st[b] (pt) | Trees removed (%) | Percent yield loss per st (pt) due to HLB | Percent tree TCA reduction due to HLB |
| US-942 | 291 | 4 | 227 (215) | 14 | 22 (26) | 2 |
| Swingle | 240 | 0 | 180 (169) | 10 | 25 (30) | 19 |
| Carrizo | 216 | 0 | 168 (163) | 19 | 22 (25) | 31 |
| US-812 | 306 | 0 | 162 (157) | 5 | 47 (49) | 19 |
| US-827 | 235 | 0 | 141 (132) | 29 | 40 (44) | 34 |

[a] Results from Wutscher and Bowman, 1999.
[b] Cumulative yield of fruit (kg) per surviving tree (st) or per planted tree (pt), through the first four harvest seasons for 'Valencia' on these rootstocks in this trial.

# REFERENCES


1. Albrecht, U., Bowman, K.D., 2011. Tolerance of the trifoliate citrus hybrid US-897 (Citrus reticulata Blanco × Poncirus trifoliata L. Raf.) to Huanglongbing. HortScience 46, 16–22.

2. Albrecht, U., Bowman, K.D., 2012. Tolerance of trifoliate citrus rootstock hybrids to Candidatus Liberibacter asiaticus. Sci. Hortic. 147, 71–80.

3. Albrecht, U., McCollum, G., Bowman, K.D., 2012. Influence of rootstock variety on Huanglongbing disease development in field-grown sweet orange (Citrus sinensis [L.] Osbeck) trees. Sci. Hortic. 138, 210–220.

4. Barry, G.H., Castle, W.S., Davies, F.S., 2004. Soluble solids accumulation in 'Valencia' sweet orange as related to rootstock selection and fruit size. J. Am. Soc. Hort. Sci. 129, 594–598.

5. Bertolini, E., Moreno, A., Capote, N., Olmos, A., de Luis, A., Vidal, E., Pérez-Panadés, J., Cambra, M., 2008. Quantitative detection of Citrus tristeza virus in plant tissues and single aphids by real-time PCR. Eur. J. Plant Pathol. 120, 177–188.

6. Bowman, K.D., 2010. Notice to Fruit Growers and Nurserymen Relative to the Naming and Release of the US-942 Citrus Rootstock. U.S. Department of Agriculture, ARS, Washington, D.C.

7. Bowman, K.D., Garnsey, S.M., 2001. A comparison of five sour orange rootstocks and their response to Citrus tristeza virus. Proc. Fla. State Hortic. Soc. 114, 73–77.

8. Bowman, K.D., Graham, J.H., Adair, R.C., 2003. Young tree growth in a flatwoods rootstock trial with Diaprepes weevil and Phytophthora diseases. Proc. Fla. State Hortic. Soc. 116, 249–251.

9. Bowman, K.D., McCollum, G., 2015. Release of US-1516, Citrus Rootstock. U.S. Department of Agriculture, ARS, Washington, D.C.

10. Bowman, K.D., McCollum, G., 2015b. Five new citrus rootstocks with improved tolerance to huanglongbing. HortScience 50, 1731–1734.

11. Bowman, K.D., Rouse, R.E., 2006. US-812 Citrus Rootstock. HortScience 41, 832–836.

12. Castle, W.S., Bowman, K.D., Baldwin, J.C., Grosser, J.W., Gmitter Jr., F.G., 2011. Rootstocks affect tree growth, yield, and juice quality of 'Marsh' grapefruit. HortScience 46, 841–848.

13. Cheema, S.S., Kapur, S.P., Chohan, J.S., 1982. Evaluation of rough lemon strains and other rootstocks against greening-disease of citrus. Sci. Hortic. 18, 71–75.

14. Duan, Y., Zhou, L., Hall, D.G., Li, W., Doddapaneni, H., Lin, H., Liu, L., Vahling, D.W., Gabriel, C.M., Williams, K.P., Dickerman, A., Sun, Y., Gottwald, T., 2009. Complete genome sequence of citrus huanglongbing bacterium, 'Candidatus Liberibacter asiaticus ' obtained through metagenomics. MPMI 22, 1011–1020.

15. Kesinger, M., 2007. Bureau of Citrus Budwood Registration Annual Report 2006–2007. Florida Department of Agriculture and Consumer Services, Winter Haven, Florida.

16. Li, W., Hartung, J.S., Levy, L., 2006. Quantitative real-time PCR for detection and identification of Candidatus Liberibacter species associated with citrus huanglongbing. J. Microbiol. Methods 66, 104–115.

17. Nariani, T.K., 1981. Integrated approach to control citrus greening disease in India. Proc. Int. Soc. Citricult. 1, 471–472.

18. Van Vuuren, S.P., Moll, J.N., 1985. Influence of the rootstocks on greening fruit symptoms. Citrus Subtrop. Fruit J. 612, 7–10.

19. Wutscher, H.K., Bowman, K.D., 1999. Performance of 'Valencia' orange on 21 rootstocks. HortScience 34, 622–624.

20. Wutscher, H.K., Hill, L.L., 1995. Performance of 'Hamlin' orange on 16 rootstocks in East-central Florida. HortScience 30, 41–43.

# CHAPTER 7

# Yield Response of Drip-Irrigated Onion under Full and Deficit Irrigation with Saline Water in Arid Regions of Tunisia

**K. Nagaz,[1] M. M. Masmoudi,[2] and N. Ben Mechlia[2]**

[1] Laboratoire d'Aridoculture et Cultures Oasiennes, Institut des Régions Arides de Médenine, 4119 Médenine, Tunisia

[2] Laboratoire de bioclimatologie, INAT, 43 Avenue Charles Nicolle, 2083 Tunis, Tunisia

## ABSTRACT

A two-year study was conducted in arid region of Tunisia to evaluate the effects of deficit irrigation regimes with saline water on soil salinity, yield, and water use efficiency of onion grown in a commercial farm on a sandy soil and drip-irrigated with water having an E C*i* of 3.6 dS/m. Irrigation treatments consisted in water replacements of accumulated E T*c* at levels of 100% (SWB-100, full irrigation), 80% (DI-80), 60% (DI-60), when the readily available water in the control treatment (SWB-100) is depleted, deficit irrigation during ripening stage (SWB100-MDI60) and farmer method corresponding to irrigation practices implemented by the local farmers. Results on onion production and soil salinization are globally coherent between the two-year experiments and show significant difference between irrigation regimes. Higher soil salinity was maintained in the root zone with DI-60 and farmer treatments than full irrigation (SWB-100). SWB100-MDI60 and DI-80 treatments resulted also in low E *Ce* values. No significant differences were observed in bulbs fresh and dry yields, bulbs number·$ha^{-1}$ and weight from the comparison between full irrigation (SWB-100) and deficit treatments (DI-80, SWB100-MDI60). DI-60 irrigation treatment caused significant reductions in the four parameters considered in comparison with SWB-100. The farmer method caused significant reductions in yield components and resulted in increase of water usage 45 and 33% in 2008 and 2009, respectively. Water use efficiency was found to vary significantly among treatments, where the highest and the lowest values were observed for DI-60 and farmer treatments, respectively. The full irrigation (SWB-100) and deficit irrigation (DI-80 and SWB100-MDI60) strategies were found to be a useful practice for scheduling onion irrigation with saline water under the arid Mediterranean conditions of southern Tunisia.

## 1. INTRODUCTION

Water scarcity is an increasingly important issue in many parts of the world. Climate change predictions of increase in temperature and decrease in rainfall mean water will become even scarcer. This is especially the case in arid regions of Tunisia subject to frequent droughts and where restricted supply of good quality water is the most important factor limiting crop production. Irrigation of a wide range of crops is expanding around shallow wells having a salinity ranging from 3 to 9 dS/m. In the absence of sufficient rainfall events used elsewhere for natural leaching, irrigated farming in arid lands is exposed to accumulation of salts in the soils. Therefore, efficient use of saline water by irrigation is becoming increasingly important, and alternative water application method such as drip may contribute substantially to attain the twin objectives of higher productivity and optimum use of water. The trend in recent years has been towards conversion of surface to drip irrigation because cost of installation has relatively decreased with the easy access to subsidized drip irrigation equipment made possible recently. Drip irrigation, currently used in arid regions of Tunisia for high-value vegetable crop production such as onions, potatoes, peppers, and lettuces, can result in considerable saving in irrigation water, thus reducing the risks of salinization if the correct management procedures are applied.

Another way to address the issue of water shortage is through development of new irrigation scheduling techniques such as deficit irrigation, which are not necessarily based on full crop water requirement. Deficit irrigation provides a means of reducing water consumption while minimizing adverse effects on yield [1–8]. In this method, the crop is exposed to a certain level of water stress either during a particular period or throughout the whole growing season. The expectation is that any yield reduction (especially in water-limiting situations) will be compensated by increased production from the additional irrigated area with the water saved by deficit irrigation [9 11]. However, the grower must have prior knowledge of the crop yield responses to deficit irrigation. Many investigations have been carried out worldwide regarding the effects of deficit irrigation on yield of mainly horticultural crops [12–20]. Other experiments with onion [21] showed that deficit irrigation throughout the growing season of onion as 50 and 75% of E $Tc$ reduced yields from full irrigation and resulted in the highest water saving and crop water use efficiency. Kumar et al. [22] investigated also the impact of deficit irrigation strategies on onion yield and water savings. They reported that applying 80 and 60% of crop water requirements resulted in yield decreases of 14 and 38% and saved 18 and 33% of irrigation water compared to full irrigation in 2 years, respectively.

Information is limited about deficit irrigation scheduling of onion especially under arid climate where onion, classified as a sensitive plant [23], is one of the important horticultural crops in arid regions of Tunisia. Productivity is usually low, and irrigation with waters having an ECi more than 3 dS/m is commonly practiced on a routine basis without scheduling and provision drainage, and it carries the danger of a rapid soil salinization because of increased salt input. Thus, there is a need to develop strategies based on irrigation restrictions during the whole growing period that may help to save water and control salinity under conditions of high evaporative demand and chronic shortages without substantially affecting yields. Therefore, various deficit irrigation strategies have been applied to onion crop. Considering this, the objective of the study was to make quantitative assessments of both salt accumulation in the soil and yield response to full and deficit irrigation strategies with saline water in order to derive an irrigation strategy that save water in drip-irrigated onion, reduce salt input, and improve water productivity under the arid Mediterranean conditions of southern Tunisia.

# 2. MATERIALS AND METHODS

## 2.1. Experimental Site and Climate

The field experiment was carried out during the growing season of 2008 and 2009, between the months of March and August, in the southern east of Tunisia in a commercial farm situated in Médenine. The climate of the region is typical of arid conditions. The average temperature and E T*o* rate during first year ranged from 11.2 to 33.1°C and 2.1 to 7.5 mm, respectively, and in the second year, these parameters were varied, respectively, from 10.7 to 32.9°C and 2.2 to 7.2 mm during the crop growing season. Rainfall during crop growing season was recorded as 38 and 47 mm in the first and second years, respectively. The soil is of a sandy type with low organic matter content. The electrical conductivity (E C*e*) values measured before planting were, respectively, 2.6 and 2.3 dS/m (0–60 cm depth of soil) for the first and second years. The total soil available water calculated between field capacity and wilting point for an assumed onion root extracting depth of 0.60 m was 75 mm.

## 2.2. Crop Management and Experimental Design

Onion seedlings (Allium cepa L.), light red local population, were transplanted on 10 March in each year with plant and row spacing of

0.2 m and 0.50 m. The experimental design consisted of completely randomized blocks with four replications and five irrigation regimes. The experimental area was used for both years and divided into four blocks with five elementary plots per block. Each elementary plot consisted of eight rows. All plots were drip irrigated with water from a well having an $ECi$ of 3.6 $dS\cdot m^{-1}$. Each dripper had a 4 $l\cdot h^{-1}$ flow rate. Water for each block passed through a water meter, gate valve, before passing through laterals placed in every onion row. A control minivalve in the lateral permits use or nonuse of the dripper line. The uniformity of emitter water application was checked by recording the time needed for the discharge to fill a vessel of known volume. The average discharge of the emitters was found to be 4 l/h, and the uniformity coefficient was more than 90% for all the blocks.

Fertilizers were supplied for the cropping periods in the same amounts; before planting, soil was spread with 9 $t\cdot ha^{-1}$ of organic manure. Nutrient supply included N, P, and K at rates of 200, 200 and 200 $kg\cdot ha^{-1}$, respectively, which were adopted from the local practices. The P and K fertilizers were applied as basal dose before planting. Nitrogen was divided and delivered with the irrigation water in all treatments during early vegetative growth.

Five distinct water treatments were applied: SWB-100 treatment irrigated when readily available water in the root zone has been depleted and plants in that treatment received 100% of accumulated crop evapotranspiration (full irrigation treatment); two additional treatments were irrigated at the same frequency as treatment SWB-100, but with quantities equal to 60 and 80% of accumulated $ETc$ (DI-60 and DI-80, continuously deficit irrigation). These treatments were identified as continuous deficit irrigation treatments. In the fourth treatment (SWB100-MDI60), water was applied from transplanting to the midseason stage, to supply fully the $ETc$ (regulated deficit irrigation). After that stage, only 60% of $ET_c$ was applied till harvest. A fifth irrigation treatment consisted of applying the producer method corresponding to irrigation practices traditionally implemented by the local farmers, that is, a fixed amount of water is supplied to the crop every 5 days from transplanting till harvest.

The crop evapotranspiration (E T$c$) was estimated for daily time step by using reference evapotranspiration (E T$o$) combined with onion crop coefficient ($Kc$). The E T$o$ was estimated from daily climatic data collected from the institute meteorological station (data not presented) by means of the FAO-56 Penman-Monteith method given in Allen et al. [24]. The onion crop coefficient ($Kc$) was computed following the recently developed FAO-56 dual crop coefficient approach, the sum soil evaporation ($Ke$), and basal crop coefficient ($Kc\ b$) reduced by any occurrence of soil water stress ($Ks$), which provides for separate calculations for transpiration and soil evaporation ($Kc = KsKc\ b + Ke$).

For irrigation scheduling, the method used was the water balance, by means of a spreadsheet program for Excel, developed according to the methodology formulated by Allen et al. [24]. The spreadsheet program estimates the day when the target soil water depletion (readily available water, RAW) for the treatment SWB-100 would be reached and the amount of irrigation water needed to replenish the soil profile to field capacity. The program calculates the soil water depletion on daily basis using the soil water balance and projects the next irrigation event based on the target depletion (30% of total available water in the root zone, 60% of TAW). The soil depth of the effective root zone is increased with the program from a minimum depth of 0.15 m at planting to a maximum of 0.60 m in direct proportion to the increase in the onion crop coefficient.

## 2.3. Measurements and Water Use Efficiency

Matured crop was harvested after 145–150 days of transplanting for both years to estimate the onion yield. Fifteen plants per row within each plot were harvested by hand to determine fresh bulbs yield, bulbs number, and bulb weight. Also, fresh bulbs were cut into small pieces and dried at 70°C to a constant weight for estimation of dry matter.

Water use efficiency (WUE) is defined as the yield obtained per unit of water consumed, whether from irrigation or total received, therefore including the precipitation. The WUE was calculated as follows: WUE ($kg \cdot ha^{-1} \cdot mm^{-1}$) = Yield ($kg \cdot ha^{-1}$)/total water received (mm) from planting to harvest; an irrigation of 75 mm applied before planting is not included in the total.

Soil samples were collected after harvest. The soil was sampled with a 4 cm auger every 15 cm to a depth of 60 cm. Samples were air-dried and ground to pass a mesh of 2 mm size and were analyzed for E *Ce*.

## 2.4. Statistical Analysis

Analysis of variance was performed to evaluate the statistical effect of irrigation treatments on onion yields and components, WUE, and soil salinity using the STATGRAPHICS Plus 5.1 (http://www.statgraphics.com/). LSD test at 5% level was used to find any significant difference between treatment means.

# 3. RESULTS AND DISCUSSION

## 3.1. Evapotranspiration Estimates and Soil Water Balance

Figure 1 shows computed *Kc*( *KsKc b*+ *Ke*) during the cropping period in 2008 and 2009. The potential *Kc* values were found to have occurred following rain or irrigation events when the soil surface layer was wetted. The *Ke*spikes represent increased evaporation when irrigation or precipitation has wetted the soil surface and has temporarily increased $ET_c$ values (Figure 2). During the initial stage, the *Ke* spikes reach values of 0.88–0.93 following wetting by rainfall. Some of the evaporation spikes were lower during this period since only fraction of the soil surface (fw = 0.3) was wetted only by irrigation. The wet soil evaporation spikes decrease as the soil surface layer dries and the value of *Ke* became zero during the growing period when the soil surface was dried.

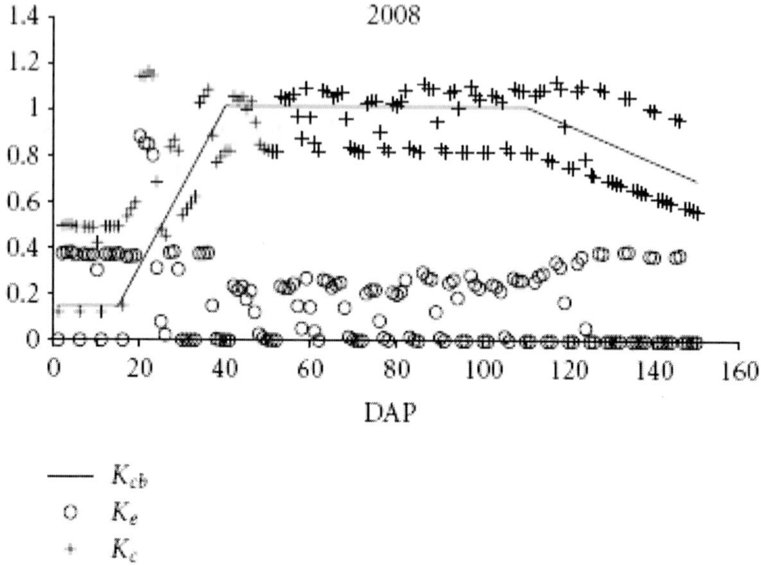


(a)

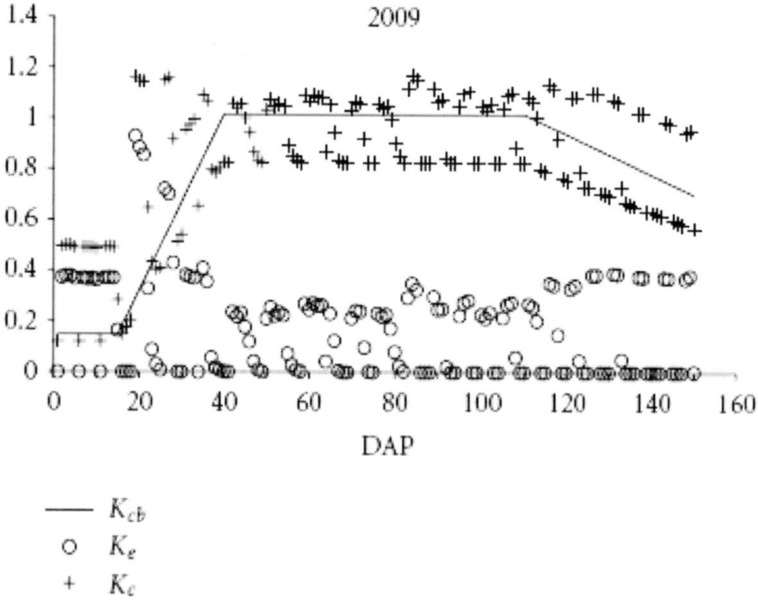


(b)

**Figure 1.** FAO 56 crop coefficient curves for onion crop during the cropping season.

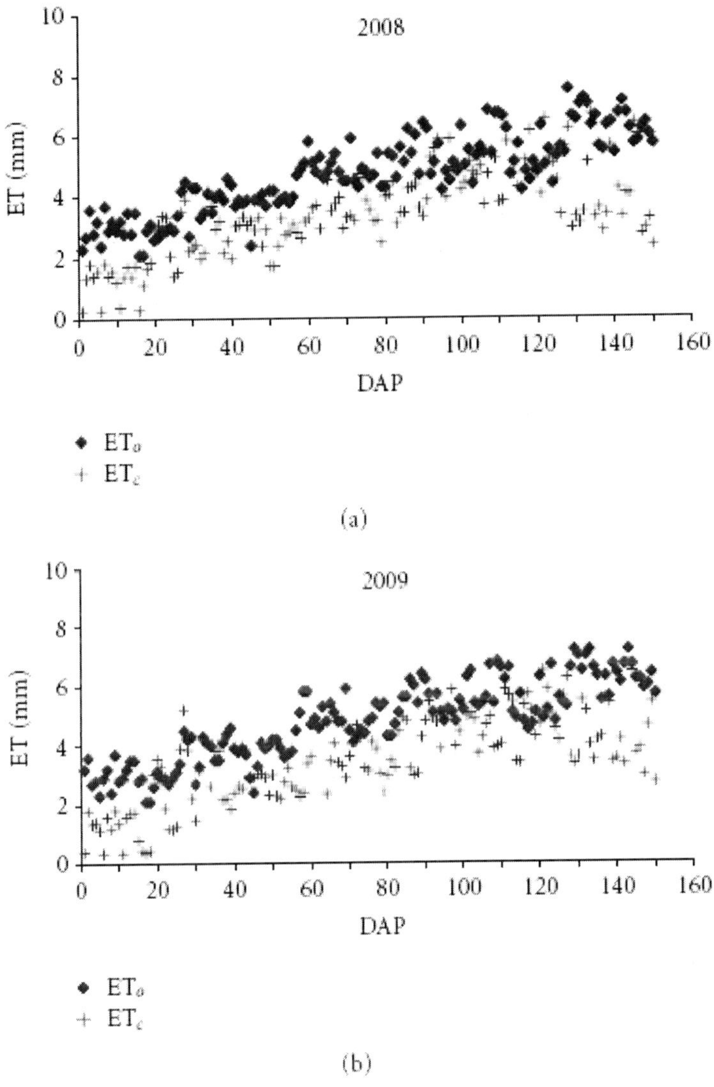


**Figure 2.** Estimated daily $ET_c$ for onion crop during the cropping season.

Figure 2 illustrates the course of daily E T*c* relative to E T*o* for onion crop. During the first 15 days after plantation, high E T*c* values were observed because of the important soil evaporation values. Most of the daily crop ET consisted of soil evaporation and controlled mainly by soil hydraulic properties and solar radiation. This period is characterized by mean values of daily E T*c* of about 1.3 and 1.24 mm, respectively, for the first and second year. As the crop canopy grew, E T*c* increased and reached its highest mean value at midseason stage (3.77 and 3.74 $mm \cdot day^{-1}$). The mean E T*c* values at the late stage were about 4.42 and 4.65 mm $day^{-1}$, respectively, for 2008 and 2009. At the late stage, where the canopy senescence began, the high E T*c* values were principally attributed to the important soil evaporation induced by the frequency of irrigation and to the high evaporative demand.

Soil Water Balance Figure 3 illustrates soil water depletion (SWD), estimated by the spreadsheet program, under SWB-100 treatment during the cropping period for the first and second year. The spreadsheet program develops a water balance and supplies information on the timing and amounts of irrigation events. This figure illustrates also the effect of an increasing root zone on the readily available water. The rate of root zone depletion at a particular moment in the season is given by the net irrigation requirement for that period. Each time the irrigation water is applied, the root zone is replenished to field capacity. Because irrigation is not applied in the spreadsheet until the soil water depletion at the end of the previous day is greater than or equal to the readily available water, occasionally plants could be subject to a slight stress on the day prior to irrigation.

## 3.2. Soil Salinity

The final average E C*e* values (0–60 cm soil depth) under the different irrigation treatments are presented in Figure 4. Initial soil salinity determined at the time of planting was 2.6 and 2.3 $dS \cdot m^{-1}$, respectively, for 2008 and 2009. After the crop harvest, E C*e* increased under all irrigation treatments in both years. However, the highest E C*e* values were found to have occurred when farmer method was used. Values of 6.4 and 6.1 dS/m were recorded, respectively, in the first and second year. The reason for the higher soil salinity obtained under farmer method may be attributed to the fact that more irrigation water under conditions of high evaporative demand would result in higher direct evaporation rates leading to an increase in salt accumulation in the soil.

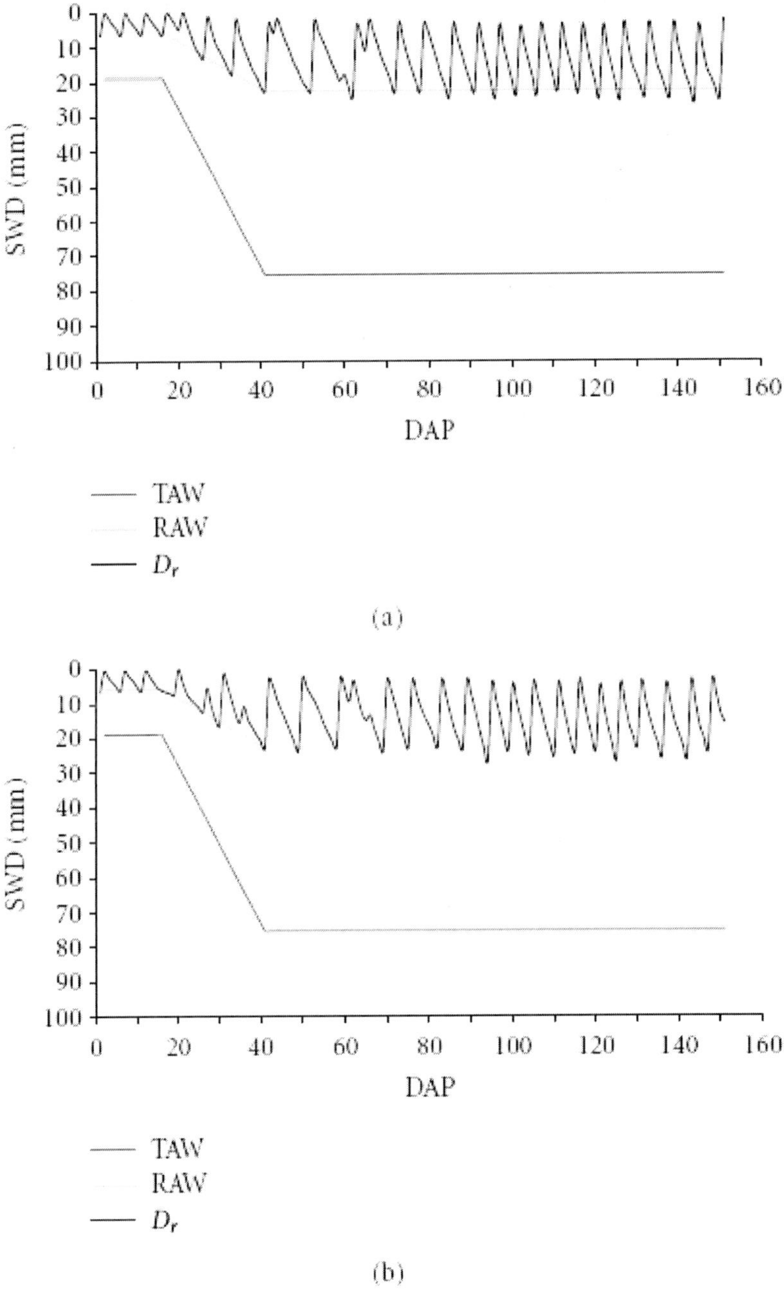


**Figure 3.** Estimated daily soil water depletion (SWD) for onion under SWB-100 irrigation treatment during the cropping season (a) 2008 and (b) 2009.

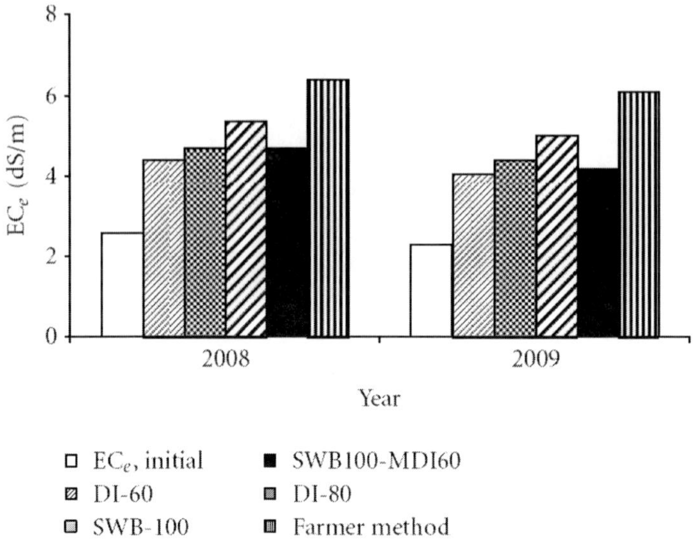


**Figure 4.** Soil salinity (E *Ce*, dS/m) under the different irrigation treatments.

With SWB irrigation scheduling (SWB-100), the average $EC_e$ value was equal to 4.4 dS/m in 2008 and to 4.05 dS/m in 2009. In both years, SWB100-MDI60 and DI-80 irrigation treatments resulted also in low E *Ce* values. The E *Ce* values are similar to the E *Ce* for SWB-100. With DI-60 irrigation treatment, E *Ce* values increased to 5.35 and $5.02\ dS \cdot m^{-1}$, respectively, in 2008 and 2009. The higher soil salinity obtained under DI-60 irrigation treatment may be attributed to little leaching of the soil expected under deficit irrigation conditions. Schoups et al. [25] reported that one consequence of reducing irrigation water use by deficit irrigation is the greater risk of increased soil salinity due to reduced leaching.

### 3.3. Yield and Yield Components

For analyzing the effect of full and deficit irrigation treatments on the final onion yield, four criteria were retained in both years: bulbs fresh yield, bulbs dry yield, bulbs number, and bulb weight. The data concerning the four parameters considered, observed for all irrigation treatments, are presented in Tables 1 and 2.

There were differences between two experiments in onion yields. Yields were highest the second year with all treatments (Table 1) because of the low initial soil salinity. SWB-100 treatment resulted in the maximum bulb fresh yields in 2008 and 2009 (24.29 and 26.10 t/ha). It was found that deficit irrigation at the fourth growth stages (SWB100-MDI60) gave nonsignificantly different yield from the full irrigation (SWB-100). Bazza [9], after conducting deficit irrigation experiment for different vegetable crops and for some cereals, concluded that stressing the crops during final stage of the growing season did not affect the crop yield significantly. Although the deficit during the last growth stage (SWB100-MDI60) was only 79 and 69 mm in the first and second years, the yield reduction was 0.79 and 0.46 t/ha. Studies conducted on irrigation of onion [12, 26–28] indicated that the bulbs fresh production of onion are highly dependent on appropriate water supply. Thus, the restrictions of irrigation water during the ripening stage according to our own experience as well as to that of other researchers [29, 30] could be implemented without leading to any quantitative or qualitative losses. Yields under the SWB100-MDI60 and SWB-100 treatments were not statistically different from that obtained with DI-80. However, a significant reduction in yields occurred with the DI-60 (Table 1). On the other hand, farmer method decreased significantly the bulb fresh yields (18.97 and 21.18 t/ha). In the study area, many farmers withhold irrigation during the last stage and provide inadequate water throughout the growing season. The farmer applies water to the crop regardless of the plant needs, and farmer seems to relate irrigation occurrences to days after planting rather than to crop growth stages progress.

Bulb dry yields increased in SWB-100 and SWB100-MDI60 treatments, and a significant decrease in bulb dry yields was observed in deficit irrigation treatment (DI-60) and farmer method (Table 1). The statistical analysis showed significant differences between the treatments except SWB-100, SWB100-MDI60, and DI-80. The variation in bulb dry yields might be attributed to variation in weight of bulb (Table 2). Olalla et al. [12] indicated that onion bulb size under drip irrigation system is directly related to amount of water applied.

**Table 1.** Yield of onion under different irrigation treatments.

| Treatments | Bulbs fresh yield (t/ha) | | | Bulbs dry yield (t/ha) | | |
|---|---|---|---|---|---|---|
| | 2008 | 2009 | Mean | 2008 | 2009 | Mean |
| SWB-100 | 24.29 | 26.10 | 25.20 | 3.045 | 3.272 | 3.16 |
| DI-80 | 23.17 | 24.98 | 24.08 | 2.905 | 3.132 | 3.02 |
| DI-60 | 22.40 | 23.44 | 22.92 | 2.808 | 2.938 | 2.87 |
| SWB100-MDI60 | 23.50 | 25.64 | 24.57 | 2.946 | 3.214 | 3.08 |
| Farmer method | 18.97 | 21.18 | 20.08 | 2.378 | 2.665 | 2.67 |
| LSD (5%) | 1.443 | 2.193 | | 0.311 | 0.280 | |

**Table 2.** Yield components under different irrigation treatments.

| Treatments | Bulbs number (1000/ha) | | | Bulb weight (g) | | |
|---|---|---|---|---|---|---|
| | 2008 | 2009 | Mean | 2008 | 2009 | Mean |
| SWB-100 | 325.50 | 335.90 | 330.7 | 74.63 | 77.70 | 76.17 |
| DI-80 | 317.70 | 330.80 | 324.3 | 72.93 | 75.50 | 74.22 |
| DI-60 | 314.70 | 317.90 | 316.3 | 71.17 | 73.64 | 72.41 |
| SWB100-MDI60 | 320.90 | 338.30 | 329.6 | 73.22 | 75.79 | 74.51 |
| Farmer method | 288.10 | 299.20 | 293.6 | 68.24 | 70.81 | 69.53 |
| LSD (5%) | 45.142 | 41.510 | | 3.017 | 4.004 | |

Bulbs number/ha and weight (Table 2) were influenced by the irrigation treatments although in both years, no significant differences in bulbs number were observed between the five treatments. However, the bulb weight for farmer method and DI-60 treatment was lowest, while SWB-100, DI-80, and SWB100-MDI60 irrigation treatments did not differ significantly from each other. Onion crop productivity is most sensitive to water stress during bulbification stage [12, 31]. Chung [31] reported that water stress during critical growth period causes reduction in bulbs number and weight. Note that the deficit irrigation treatment (DI-60) and farmer method result in higher salinity in the rooting zone than the DI-80, SWB-100, and SWB100-MDI60 strategies (Figure4). The higher salinity associated with deficit irrigation (DI-60) and farmer methods were sufficient to cause reduction in onion yield, through a reduction in bulbs number and weight (Table 2). These results obtained under the prevailing climatic conditions support the practicality of the DI-80, SWB-100, and SWB100-MDI60 strategies to facilitate the use of saline water for irrigation of onion.

## 3.4. Water Use Efficiency

Data on the amounts of applied irrigation water under the different irrigation treatments in both years are presented in Table 3. For all treatments, total water supply ranged from about 332 to 750 mm. The irrigation water applied to SWB-100, DI-80, DI-60, SWB100-MDI60, and farmer method were, respectively, 492, 394, 295, 413, and 712 mm for the first year and 474, 379, 285, 405, and 631 mm in the second year. The amounts of irrigation water in the two experiments were similar to those reported by Kadayifci et al. [32], Mermoud et al. [33], Kumar et al. [22], and Bekele and Tilahun [21]. Among the irrigation treatments tested, the irrigation water supply was the highest under farmer method in both years. The full irrigation (SWB-100) received 45 and 33.1% less irrigation water than the farmer method, respectively, in 2008 and 2009 (Table 3). Therefore, irrigation water applied to farmer method was considered in excess of actual irrigation water requirement. We thus used full irrigation treatment (SWB-100), not the farmer method, as the basis for comparison of irrigation treatments in saving water. As expected, the deficit irrigation treatments required less water than the full-irrigated treatment (SWB-100). The net saving in irrigation water with DI-80, DI-60, and SWB100-MDI60 were, respectively, 19.9, 40, and 16.1% for the first year and 20, 39.9, and 14.6% in the second year when compared with full irrigation treatment (SWB-100).

Water use efficiency (WUE) is defined as the relationship between units produced and volume of irrigation water applied [34]. Table 4 shows the WUE value in our experiment expressed as kilograms of fresh bulb weight produced per cubic meter of irrigation and total water received from planting to harvest. The WUEs values obtained for the first and second year experiments are comparable with those obtained in other field studies [21, 22, 32] and were affected by irrigation treatments.

For both experiments, the WUE with SWB-100 treatment was not significantly different from those obtained with DI-80 and SWB100-MDI60 treatments but statistically different from that obtained with DI-60 and farmer treatments. The difference was also significant between DI-80 treatment and the DI-60 and farmer treatments. These last two show a statistical difference between them.

The lowest IWUE was achieved under the farmer irrigation method where the grower used no quantitative assessment to determine the amount of irrigation water to be applied. The low WUE for the farmer method during the two experiments can be attributed to reduced yields but also to higher irrigation water use. The deficit irrigation treatments with 20% and 40% less irrigation water (DI-80, DI-60) gave proportionally higher IWUE compared to other treatments (Table 3). Our results therefore suggest that irrigation water requirements of onion crop can easily be reduced by 42–45% and 36–40% of farmer method, respectively, in 2008 and 2009, without significant yield reduction in the Mediterranean arid regions if the SWB100-MDI and DI-80 deficit irrigation practices are adopted. This suggests that increasing the irrigated area with the saved water would compensate for any yield loss.

**Table 3.** Water supply (mm) under different irrigation treatments.

| Treatments | Irrigation (I)* | Irrigation water saved (%) | Precipitation (P) | (I + P) |
|---|---|---|---|---|
| | | 2008 | | |
| SWB-100 | 492 | — | 38 | 530 |
| DI-80 | 394 | 19.9 | 38 | 432 |
| DI-60 | 295 | 40.0 | 38 | 333 |
| SWB100-MDI60 | 413 | 16.1 | 38 | 451 |
| Farmer method | 712 | — | 38 | 750 |
| | | 2009 | | |
| SWB-100 | 474 | — | 47 | 521 |
| DI-80 | 379 | 20.0 | 47 | 426 |
| DI-60 | 285 | 39.9 | 47 | 332 |
| SWB100-MDI60 | 405 | 14.6 | 47 | 452 |
| Farmer method | 631 | — | 47 | 678 |

* An irrigation of 75 mm supplied just before planting is not included in these totals.

**Table 4.** Water use efficiency (kg·m$^{-3}$) under different irrigation treatments.

| Treatments | IWUE | | | TWUE | | |
|---|---|---|---|---|---|---|
| | 2008 | 2009 | Mean | 2008 | 2009 | Mean |
| SWB-100 | 4.94 | 5.51 | 5,23 | 4.58 | 5.01 | 4,80 |
| DI-80 | 5.88 | 6.59 | 6,24 | 5.36 | 5.86 | 5,61 |
| DI-60 | 7.59 | 8.23 | 7,91 | 6.73 | 7.06 | 6,90 |
| SWB100-MDI60 | 5.69 | 6.33 | 6,01 | 5.21 | 5.67 | 5,44 |
| Farmer method | 2.66 | 3.36 | 3,01 | 2.53 | 3.12 | 2,83 |
| LSD (5%) | 0.881 | 0.946 | | 1.175 | 1.179 | |

# 4. CONCLUSION

Results of this study indicate that full irrigation (SWB-100) and deficit irrigation (SWB100-MDI60 and DI-80) treatments decreased the soil salinity. Higher soil salinity was maintained in the root zone with DI-60 deficit irrigation strategy and farmer method. Onion yields were influenced by irrigation treatments. Bulb fresh yields of deficit-irrigated treatment (DI-60) and farmer method were significantly lower than those in full irrigation treatment (SWB-100). Treatments SWB100-MDI60 and DI-80 give also good yields. Deficit-irrigation treatment DI-60 resulted in lower yields and in higher salinity in the root zone than full irrigation (SWB-100) and SWB100-MDI60 and DI-80 treatments. The "fixed amount approach" used by the farmer gave the lowest bulb fresh yields, that is, 22–19% less than SWB-100 with 45 and 33% more water and resulted in higher salinity in the rooting zone.

The water use efficiency (WUE) for onion yield was significantly affected by irrigation treatments. The lowest values occurred under the farmer method, while the highest values were obtained under DI-60 deficit-irrigation treatment. Although high efficiencies were observed for the most severe restriction (DI-60), the yield and quality obtained under this treatment do not allow opting for such important reduction. The relatively high yields and water use efficiency values noted under DI-80 and SWB100-MDI60 treatments indicate the high potential of the onion crop to valorize irrigation waters of limited quality under mild continuous and regulated deficit-irrigation.

Based on results, obtained under actual farming conditions, it can be concluded that the full irrigation (SWB-100) and continuous and regulated deficit-irrigation (DI-80 and SWB100-MDI60) strategies offer significant advantage for both onion yields and WUE and reduce the build-up of salinity compared to the DI-60 and farmer irrigation practices in onion production under arid conditions. As a result of this research, full irrigation based on SWB method (SWB-100) is recommended for irrigation of onion crop under the arid climate of southern Tunisia. In case of situations where water supply is limited, irrigation of onion could be scheduled using DI-80 and SWB100-MDI60 continuously and regulated deficit-irrigation strategies.

# REFERENCES


1. M. English and B. Nakamura, "Effects of deficit irrigation and irrigation frequency on wheat yields,"Journal of Irrigation and Drainage Engineering, vol. 115, no. 2, pp. 172–184, 1989.

2. M. English and S. N. Raja, "Perspectives on deficit irrigation," Agricultural Water Management, vol. 32, no. 1, pp. 1–14, 1996.

3. F. T. Mugabe and E. Z. Nyakatawa, "Effect of deficit irrigation on wheat and opportunities of growing wheat on residual soil moisture in Southeast Zimbabwe," Agricultural Water Management, vol. 46, no. 2, pp. 111–119, 2000.

4. G. Ghinassi and P. Trucchi, "Deficit irrigation trials on maize in an Mediterranean semi-arid environment," International Water and Irrigation, vol. 21, no. 1, pp. 12–17, 2001.

5. C. R. Kirda, "Deficit irrigation scheduling based on plant growth stages showing water stress tolerance,"Deficit Irrigation Practices, FAO Water Reports, vol. 22, pp. 3–10, 2002.

6. X. Mao, M. Liu, X. Wang, C. Liu, Z. Hou, and J. Shi, "Effects of deficit irrigation on yield and water use of greenhouse grown cucumber in the North China Plain," Agricultural Water Management, vol. 61, no. 3, pp. 219–228, 2003.

7. R. K. Panda, S. K. Behera, and P. S. Kashyap, "Effective management of irrigation water for wheat under stressed conditions," Agricultural Water Management, vol. 63, no. 1, pp. 37–56, 2003.

8. Y. Zhang, E. Kendy, Y. Qiang, L. Changming, S. Yanjun, and S. Hongyong, "Effect of soil water deficit on evapotranspiration, crop yield, and water use efficiency in the North China Plain," Agricultural Water Management, vol. 64, no. 2, pp. 107–122, 2004.

9. M. Bazza, "Improving irrigation management practices with water-deficit irrigation," in Crop Yield Response to Deficit Irrigation, Developments in Plant and Soil Sciences, C. Kirda, P. Moutonner, C.

Hera, and D. R. Nielsen, Eds., vol. 84, pp. 49–71, Kluwer Academic, Dodrecht, The Netherlands, 1999.

10. H. Kirnak, I. Tas, C. Kaya, and D. Higgs, "Effects of deficit irrigation on growth, yield, and fruit quality of eggplant under semi-arid conditions," Australian Journal of Agricultural Research, vol. 53, no. 12, pp. 1367–1373, 2002.

11. M. H. Ali, M. R. Hoque, A. A. Hassan, and A. Khair, "Effects of deficit irrigation on yield, water productivity, and economic returns of wheat," Agricultural Water Management, vol. 92, no. 3, pp. 151–161, 2007.

12. F. J. Olalla, A. Domínguez-Padilla, and R. López, "Production and quality of the onion crop (Allium cepaL.) cultivated under controlled deficit irrigation conditions in a semi-arid climate," Agricultural Water Management, vol. 68, no. 1, pp. 77–89, 2004.

13. F. J. Olalla, J. A. de Juan, and C. Fabeiro, "Growth and production of onion crop (Allium cepa L.) under different irrigation schedulings," European Journal of Agronomy, vol. 3, no. 1, pp. 85–92, 1994.

14. F. J. Olalla, J. A. de Juan, and C. Fabeiro, "Growth and yield analysis of soybean (Glycine Max Merr.) under different irrigation schedules in Castilla-La Mancha," European Journal of Agronomy, vol. 3, no. 3, pp. 187–196, 1994.

15. M. Bazza and M. Tayaa, "Contribution to improve sugar beet deficit irrigation," in Crop Yield Response to Deficit Irrigation, Developments in Plant and Soil Sciences, C. Kirda, P. Moutonner, C. Hera, and D. R. Nielsen, Eds., vol. 84, pp. 72–97, Kluwer Academic, Dodrecht, The Netherlands, 1999.

16. C. Fabeiro, F. J. Olalla, and J. A. de Juan, "Yield and size of deficit irrigated potatoes," Agricultural Water Management, vol. 48, no. 3, pp. 255–266, 2001.

17. C. Fabeiro Cortés, F. J. Olalla, and J. A. de Juan, "Production of muskmelon (Cucumis melo L.) under controlled deficit irrigation in

a semi-arid climate," Agricultural Water Management, vol. 54, no. 2, pp. 93–105, 2002.

18. C. Fabeiro Cortés, F. J. Olalla, and R. López Urrea, "Production of garlic (Allium sativum L.) under controlled deficit irrigation in a semi-arid climate," Agricultural Water Management, vol. 59, no. 2, pp. 155–167, 2003.

19. C. Fabeiro Cortés, F. J. Olalla, R. López Urrea, and A. Domínguez, "Production and quality of the sugar beet (Beta vulgaris L.) cultivated under controlled deficit irrigation conditions in a semi-arid climate,"Agricultural Water Management, vol. 62, no. 3, pp. 215–227, 2003.

20. S. M. Sezen, A. Yazar, A. Akyildiz, H. Y. Dasgan, and B. Gencel, "Yield and quality response of drip irrigated green beans under full and deficit irrigation," Scientia Horticulturae, vol. 117, no. 2, pp. 95–102, 2008.

21. S. Bekele and K. Tilahun, "Regulated deficit irrigation scheduling of onion in a semiarid region of Ethiopia," Agricultural Water Management, vol. 89, no. 1-2, pp. 148–152, 2007.

22. S. Kumar, M. Imtiyaz, A. Kumar, and R. Singh, "Response of onion (Allium cepa L.) to different levels of irrigation water," Agricultural Water Management, vol. 89, no. 1-2, pp. 161–166, 2007.

23. E. V. Maas and G. J. Hoffman, "Crop salt tolerance: current assessment," Journal of the Irrigation and Drainage Division, vol. 103, no. 2, pp. 115–134, 1977.

24. R. G. Allen, L. S. Perreira, D. Raes, and M. Smith, Crop Evapotranspiration: Guidelines for Computing Crop Water Requirements, FAO Irrigation and Drainage, Rome, Italy, 1998.

25. G. Schoups, J. W. Hopmans, C. A. Young et al., "Sustainability of irrigated agriculture in the San Joaquin Valley, California," Proceedings of the National Academy of Sciences of the United States of America, vol. 102, no. 43, pp. 15352–15356, 2005.

26. S. O. Koriem, M. M. A. El-Koliey, and M. F. Wahba, "Onion bulb production from "Shandwell" sets as affected by soil moisture stress," Assuit Journal of Agriculture Science, vol. 25, pp. 185–193, 1994.

27. C. C. Shoke, E. B. G. Feibert, and L. D. Saunders, "Onion yield and quality affected by soil water potential as irrigation threshold," Horticulture Science, vol. 33, no. 7, pp. 1188–1191, 1998.

28. C. C. Shoke, E. B. G. Feibert, and L. D. Saunders, "Irrigation criteria for drip-irrigated onions,"Horticulture Science, vol. 35, no. 1, pp. 63–66, 2000.

29. W. R. Jordan, "Whole plants response to water deficit: an overview," in Limitations of Efficient Water Use in Crop Production, H. M. Taylor, W. R. Jordan, and T. R. Sinclair, Eds., pp. 289–317, ASA, CSSA and SSSA, Madison, Wis, USA, 1983.

30. T. A. Howell, R. H. Cuenca, and K. H. Solomon, "Crop yield response," in Management of Farm Irrigation System, G. J. Hoffman, Ed., vol. 9, pp. 93–122, Monograph, St. Joseph, Mich, USA, 1990.

31. B. Chung, "Irrigation and bulb onion quality," Acta Horticulturae, vol. 247, pp. 233–236, 1989.

32. A. Kadayifci, G. I. Tuylu, Y. Ucar, and B. Cakmak, "Crop water use of onion (Allium cepa L.) in Turkey,"Agricultural Water Management, vol. 72, no. 1, pp. 59–68, 2005.

33. A. Mermoud, T. D. Tamini, and H. Yacouba, "Impacts of different irrigation schedules on the water balance components of an onion crop in a semi-arid zone," Agricultural Water Management, vol. 77, no. 1–3, pp. 282–293, 2005.

34. P. Steduto, "Water use efficiency," in Sustainability of Irrigated Agriculture, L. S. Pereira, R. A. Feddes, J. R. Gilley, and B. Lesaffre, Eds., vol. 312 of NATA ASI Series E: Applied Sciences, Kluwer Academic, London, UK, 1996.

# CHAPTER 8

# Water catchment with native species in the Brazilian Semiarid

**João Vianey Fernandes Pimentel[1], Hugo Orlando Carvallo Guerra[2], Francisco Jardel Rodrigues da Paixão[3], Luciano F. Marques[4], Leandro do Vale[4], Francisca N. Brito[5]**

[1]Agroecology and Environment of Federal Institute of Education Science and Technology of Rio Grande do Norte (UFRN), Quixeramobim, Brazil

[2]Agricultural Engineering, Department of Graduate Agricultural Engineering, Federal University of Campina Grande (DEAG/ UFCG), Campina Grande, Brazil

[3]Technological Axis Natural Resources Centec Institute/Fatec Sertão Central, Quixeramobim, Brazil; Jardel.

[4]State University of Maranhão (UEMA), Balsas, Brazil

[5]Agribusiness Centec Institute/FATEC Sertão Central, Undergraduate Scholarship Foundation Protection Research of Ceará, FUNCAP, Quixeramobim, Brazil

## ABSTRACT

Xerophyticplants are well adapted to the semiarid conditions because they have a high potential for the use of rainwater. They can store it in their roots (xylopodiums). The Amburanacearensis, commonly called Cumaru, native of the Brazilian semiarid, is very important because of its multiple-use: good quality wood, active food ingredient, perfume and medicine production. However, its management is practically unknown. Aiming to know more about this culture, an experiment was conducted to quantify the Cumaruwater needs, evaluate its water-use efficiency and the capacity of roots to store water. Five levels of water (0%, 25%, 50%, 75% and 100% of evapotranspiration) and five-plant spacing (0.20 × 0.20 m, 0.20 × 0.10 m, 0.20 × 0.05 m, 0.10 × 0.10 m and 0.05 × 0.05 m) were tested, and the results obtained analyzed statistically throughout F test. No influence of soil water level and plant spacing in the moisture content of the whole plant, in the aerial part or in the xylopodiums was observed. However, there was an effect of the treatments on the water stored in the whole plant in the aerial part or in the roots, per unit area. More than 80% of the water was stored in the roots. The crop growing with the smaller spacing without irrigation was shown to be quite efficient in using the rainwater, compared with traditional cultures irrigated.

## KEYWORDS

Xerophytic; Xylopodium; Semiarid; Water Use Efficient

## 1. INTRODUCTION

Duke states that one of the ways to resist the pauperism in the semiarid regions is to eliminate the uncertainness of annual tillage due to the periodic wet and dry weather conditions, thus giving the farmer a safe agricultural based on the use of drought-resistant plants which provides food for the creation of cattle with plenty of fodders, and way of life more methodical. The use of xerophytes, particularly those presenting xylopodiums, seems to offer an adequate alternative for the agriculture.

Gnadlinger [2] indicates that xerophyliticplants of the caatinga region ("White Wood") are well adapted to dry conditions. They are leafless and take a white coloration during the dry season to reduce water losses through transpiration. Cacti, Umbutree (Spatodiatuberosa), Juazeiro (Zizyphusjoazeiro), Waxpalm (Coperniciacerifera) and many other xerophyliticspecies have their own mechanisms of water absorption and storage and prevention against perspiration losses. The roots of the cactus, for example, near the surface, absorb rain water quickly; Umbu trees store water in the roots (until 2500 kg/tree). Duke [3] confirms that the roots of droughtresistant plants have an anatomy and physiology rigorously synchronized with the ecology of the region and they are much more important as the organs that were considered.

The Amburanacearensis also known as cumaru, a native plant of multiple uses with its good quality timber, easy to work with and pleasant aroma, is sold comercially under the name Cherry. Its roots (xilopódios), bark and seeds, produce coumarin, active principle which, apart from being used in food industries (candies and cookies), cigarettes and tobacco, perfume industries as fixative is also used in the production of drugs such as the Cumaru syrup or homemade lickers, of broad popular use, proven scientifically as anti-inflammatory and bronchodilator [4]. For its beauty, the tree can be used as ornamental landscape projects. For the soil recovery and the restoration of degraded environments, it is used both at the initial and the later stages of reforestation, even as riparian vegetation in areas with periodic and short duration floodings [5]. In agroforestry systems, the cumaru can be used as windbreaks and tracks on tree plantations. As forage, its leaves and pods are eaten by goats, both green and dry, and by the cattle after drying. It is also of great importance for beekeeping and meliponiculture, providing nectar in the dry season, ranking among the 18 most used species by native bees for pollen and/or nectar and as nesting sites, besides the use of its wood in the construction of hives [6].

It already studies the possibility of plant cultivation in flower beds to produce herbals of broad popular use. Canuto [4] indicates that young plants, from 7 months old, may replace the mature plants husks, reducing the risk of extinction by vegetation exploitation.

Considering that most of the rain water in the semiarid is lost without any productive use, the cultivation of native species, especially those that store water in the xylopodiums, constitutes an alternative to capture the water, contributing to human sustainability. The objective of this study, thus, was to quantify the water needs of Cumaru in flower beds, determining how much of this water is stored in the xilopodiums and its water-use efficiency.

## 2. MATERIAL AND METHODS

The experiment was conducted at the Center for Environmental Education (NEA), of the Brazilian Institute of Environment (IBAMA), at Quixeramobim, one of the largest municipalities of Central Ceará State, with an area of 3275 $km^2$. It is located in the semiarid regionwith geographic coordinates with 5th 12'08" south latitude, 39°17'11" west longitude and altitude of 196 m above sea level, 204 km from Fortaleza, capital of Ceará. According to IBGE [7], the total population was 71,887 inhabitants, including rural and urban areas. The municipality of Quixeramobim presents prevalence of soils Bruno No Calcite (37.4%), Planossolo Solodic (34.7%), Litholic (9.6%), Brunizem Reddish (8.8%), Red-Yellow Podzolic (6.6%), Vertisol (2.4%) and Regossolo (0.5%), Bruno No Calcite (37.4%), Planossolo Solodic (34.7%), Litholic (9.6%), Brunizem Reddish (8.8%), Red-Yellow Podzolic (6.6%), Vertisol (2.4%) and Regossolo (0.5%), with slopes of 0% - 5%. The climate, according to Köppen classification, is the type BShw'-semiarid with short rainy season in summer-autumn, with concentration of rainfall in the months of March and April. The normal average rainfall of the municipality is 708.5 mm [8]. The average temperature ranges between 22.8°C and 32.5°C. The main water resources are the Fogareiro reservoir with a capacity of 118,820,000 $m^3$, perpetuating the river and the Quixeramobimdam with a capacity of 54,000,000 $m^3$, in addition to various public and private dams of medium and small size, without counting the lakes and groundwater reserves of crystalline aquifers.

Five flowerbeds 1 m wide, 15 m long, and 40 cm height were excavated, internally lined with plastic sheeting to a depth of 20 cm and filled with a mix of soil and manure in ratio of 100 liters of manure per square meter of bed, corresponding to the proportion of 25% by volume. Planting took place with two seeds per hill plot, according to the spacing provided for each treatment.

Five water contents were tested: 0%, 25%, 50%, 75% and 100% of the evapotranspiration. The 0% treatment received only the natural precipitation. The other treatments were obtained complementing the precipitation with irrigation water supplied with three drip tapes with drips every 30 cm. The flower bed was divided in five parcels of 1.0 $m^2$ each one with a different spacing: 0.20 × 0.20 m; 0.20 × 0.10 m; 0.20 × 0.05 m; 0 10 × 0.10 m and 0.05 × 0.05 m.

After stabilization of the stand at 90 days (taking 30 days to complete germination of all seeds and the remainder to reach uniformity), with thinning and replanting of plants the biometry was conducted collecting monthly a plant for square meter, in which the total, roots and aerial fresh weight and the total, root and shoot dry matter weight were measured. In the present work, only the data obtained at the end of the experiment (210 DAS) is presented. The experiment was conducted in the period from January to August 2010.

During the experimental period, from January to August 2010, the monthly precipitation registered in the area was:

| Month | Precipitation (mm) |
|---|---|
| January | 76.0 |
| February | 19.0 |
| March | 8.5 |
| April | 235.0 |
| May | 62.5 |
| June | 133.0 |
| July | 0.0 |
| August | 0.0 |
| **Total** | **613.0** |

The evapotranspiration used was the actual or real, determinated by the water consumption of the plant in the flowerbeds that received an irrigation depth of 100% of the evapotranspiration (Treatment I5), subtracting the drained from the applied water, a day before the irrigation.

The total fresh weight, roots and aerial part was measured on a balance with a scale accuracy of 0.1 g. For the determination of dry matter, the parts of the plant, total, root and shoot were placed in paper bags, dried in a oven with controlled temperature of 65°C for 24 hours and weighed on a balance with a scale accuracy of 0.1 g.

The moisture content was determined by the ratio beween the difference of fresh and dry weights and fresh t weight, multiplied by one hundred, (fresh-dry/fresh) × 100. The moisture content in each plant per square meter of flowerbed, was obtained multiplying the moisture content of the fresh plant weight by the number of plants per square meter.

The water use efficiency was determined using the relationship between the total cumaru fresh weight produced and the amount of water consumed per $m^2$, expressed in $g \cdot L^{-1}$.

The statistical design was a randomized block, split plot distributed, with five water levels (I1 = precipitation only; I2 = precipitation + irrigation to reach 25% of ET; I3 = precipitation + irrigation to reach 50% of ET; I4 = precipitation + irrigation to reach 75% of ET and I5 = precipitation + irrigation to reach 100% of ET) and five spacings (S1 = 0.20 × 0.20 m, S2: 0.20 × 0.10 m; S3: 0.20 × 0, 0.5 m; S4: 0.10 × 0.10 m, and S5: 0.05 × 0.05 m) and three replications, totalizing 75 plots.

The effects of treatments were assessed by variance and regression analysis, using the software SISVAR, DEX/ UFLA [9].

## 3. RESULTS AND DISCUSSION

It is observed on Table 1 that there was no significant effect of the water levels, or the plant spacing or the interaction on the moisture content of the whole plant, xilophodiumor aerial part of the young plants of Cumaru beds, at 210 days after sowing. This indicates that the Cumaruis a rustic plant, adapted to low soil water contentconditions, and that supplementary irrigation does not increase the moisture content of the plant, even in the water reserve organs (xylopodiums).

**Table 1**. Analysis of variance and means for the moisture in the whole plant (% UT), in the xilopódios (UX%), in the aerial part (UPA%) and water content in the whole plant (CAP), in the xylopodiums (CAX) and in the aerial part (CAPA), with respect to the water applied (I) and plant spacing (S) in young plants of Cumaru, 210 days after sowing, Quixeramobim, CE, 2010.

| F.V. | G.L. | Medium squares | | | | | |
|---|---|---|---|---|---|---|---|
| | | % UT | % UX | % UPA | CAP | CAX | CAPA |
| **Water (I)** | 4 | 18.99 ns | 31.83 ns | 69.80 ns | 0.19 ns | 0.11 ns | 0.03 ns |
| **Spacing (S)** | 4 | 113.87 ns | 41.78 ns | 70.20 ns | $1.37^{**}$ | $1.09^{**}$ | $0.08^{*}$ |
| **- Linear regression** | 1 | | | | $55.80^{**}$ | $42.41^{**}$ | $0.24^{**}$ |
| **- Quadratic regression** | 1 | | | | 2.60 ns | 3.37 ns | 0.00 ns |
| **Water × Spacing (Ix S)** | 16 | 73.06 ns | 53.73 ns | 152.95 ns | 0.19 ns | 0.15 ns | 0.03 ns |
| **Blocks** | 2 | 144.98 ns | 86.43 ns | 509.59 ns | 0.51 ns | 0.42 ns | 0.03 ns |
| **CV(%) (water)** | | 9.21 | 7.00 | 18.52 | 25.58 | 22.77 | 13.19 |
| **CV(%) (spacing)** | | 9.40 | 8.85 | 9.51 | 15.30 | 14.58 | 11.58 |
| **CV(%) (interaction)** | | 10.74 | 9.81 | 15.78 | 25.23 | 22.97 | 14.52 |
| **Water applied (I)** | | Mean without transformation | | | | | |
| | | UT | UX | UPA | CAP | CAX | CAPA |
| | | (%) | (%) | % | $(L \cdot m^{-2})$ | $(L \cdot m^{-2})$ | $(L \cdot m^{-2})$ |
| I1(0.00 ET) | | 74.25 | 77.10 | 64.63 | 1.24 | 1.11 | 0.15 |
| I2(0.25 ET) | | 73.26 | 77.13 | 63.09 | 1.21 | 1.07 | 0.16 |
| I3(0. 50 ET) | | 76.06 | 77.88 | 67.94 | 2.18 | 1.74 | 0.46 |
| I4(0.75 ET) | | 73.34 | 80.55 | 62.81 | 1.15 | 1.06 | 0.14 |
| I5(1.00 ET) | | 74.25 | 77.35 | 63.05 | 1.59 | 1.29 | 0.31 |
| **Spacing (S)** | | | | | | | |
| S1 (0.20 × 0.20 m) | | 76.88 | 80.72 | 66.72 | 0.45 | 0.39 | 0.06 |
| S2 (0.20 × 0.10 m) | | 70.06 | 76.40 | 61.15 | 0.65 | 0.59 | 0.08 |
| S3 (0.20 × 0.05 m) | | 75.23 | 77.55 | 65.51 | 1.72 | 1.33 | 0.41 |
| S4 (0.10 × 0.10 m) | | 76.04 | 78.30 | 64.91 | 1.45 | 1.23 | 0.23 |
| S5 (0.05 × 0.05 m) | | 72.94 | 77.06 | 63.23 | 3.10 | 2.74 | 0.44 |

**Significant at the 1% level of probability ($p < 0.01$); *Significant at 5% level of probability ($p < 0.05$); ns: not significant.

Chaves Filho & Stacciarini-Seraphin [10] studying a savannah native specie, Solanumlycocarpum (commonly known as lobeira), found no difference in relative water content of leaves when submitted to water stress treatments. However, they claim that it cannot be conclude that these plants are in fact insensitive to marked dehydration of the upper layers of the soil during the dry season. Everything indicates that these plants are physiologically adapted to the conditions of "dry season" invariably associated to the existence of savannas, as the "caatinga". Fahn & Cutler [11] said that plants, established from the beginning in environments with restricted water availability developed anatomic and physiological adaptations, which resulted in the adaptation success to the hostile conditions of water that these environments offer.

The water content of plants per square meter of bed for the whole plants, xylopodiums and shoots was not affected by the soil water content; however it was significantly affected by plant spacing (Table 1). Obviously, as the spacing among plants decrease, the number of plants per square meter increases, increasing also the water catch per unit area. Table 1 shows that, as the culture become denser, the uptake of water per square meterincreased, capturing 3.10 $L \cdot m^{-2}$ of bed with the shortest plant spacing tested (0.05 × 0.05 m) compared with the 0.45 $L \cdot m^{-2}$ with the plant spacing of 0.20 × 0.20 m. It is apparent that between S3 and S4 treatments the results were similar, because even though they are two different spatial arrangements (0.20 × 0.05 m and 0.10 × 0.10 m), the area is the same (0.01 $m^2$). Approximately 88.4% of the water was accumulated on the xylophodiums.

As there was not response of the cumaru to the irrigation, the water use efficiency was calculated only for the treatment that not received irrigation (0.00 ET) and only received rainwater. Similarly, as the most adequate spacing recommended from Pimentel [12] was the densest, it was considered only the treatment S5 (0.05 × 0.05 m).For this treatment, the water-use efficiency obtained was For this treatment, the water-use efficiency obtained was 10.02 $g \cdot L^{-1}$. The water-use efficiency obtained in the present study is superior to those found on the references. Silva et al. [13] indicate that cultures with great efficiency in water use, such as irrigated sugarcane forage has indexes of 1.67 $g \cdot L^{-1}$. Barros Júnior et al. [14] obtained in treatments with 100% of available water in the soil, water efficiency index of 2.80 $g \cdot L^{-1}$ for castor beans, considered a water resistant crop. The rosemary-pepper, traditional medicinal plant in the semiarid that produces an essential oil, had a water use efficiency of 1.26 $g \cdot L^{-1}$ with an irrigation depth of 1.65 of the reference evapotranspiration [15].

Attention should be paid that, the cumaru, with seven months, has the potential to be fully used (root, stem and leaves) in the pharmaceutical industry. Because, at this age, it already contains the active principles in all parts [4]. It can be supplied, fresh, directly by the farmer since the moisture content for the manipulation (between 8 and 14%) can be obtained by the proper pharmaceutical industry.

# 4. CONCLUSIONS

1) The moisture content in the whole plant, in xylopodiums and aerial part of the plant, cultivated in flower beds, were indifferent to the use of supplementary irrigation and spacing.

2) The water content of the whole plant, xylophodiums and aerial part of the plant per $m^2$ of bed was not influenced by the soil water content, however, increased with the decrease of plant spacing which meant increase of plants per $m^2$.

3) The highest water use efficiency was obtained with the lowest plant spacing.

4) It is recommended that further studies will be conducted with the use of native plants in the semiarid. Besides the socioeconomically importance, they are a great alternative to use the scarce water resources in this region, ensuring thus the ecological sustainability.

## REFERENCES


1. Duke, J.G. (2004) The Northeast and crops xerófilas. 4th Edition, Bank of Northeast of Brazil, Fortaleza.

2. Gnadlinger, J. (2003) Catchment and management of rainwater and sustainable development of semi-arid Brazil—An integrated vision. 4th Brazilian Symposium on Capture and Management Rainwater, Petrolina, July 2003.

3. Duke, J.G. (2004) Soil and water in the drought polygon. 6th Edition, Bank of Northeast of Brazil, Fortaleza.

4. Canuto, K.M. (2007) Chemical aspects of the interdisciplinary study (agronomy chemistry-pharmacology) of Amburanacearensis A. C. Smith. 313 f. Ph.D. Thesis, Federal University of Ceará, Fortaleza.

5. Maia, G.N. (2004) Caatinga: Trees and shrubs and their utilities. Reading and Art, Sao Paulo.

6. Marine, I.V., Freitas, M.F., Zanella, F.C.V., Caldas, A.L. (2002) Plant species of Caatinga used by indigenous stingless bees as font of resources and nesting place. 1st Brazilian Congress of University Extension, University Press, Joao Pessoa.

7. Brazilian Institute of Geography and Statistics (IBGE) (2010) Synopsis of the demographic census. http://www.censo2010.ibge.gov.br/sinopse/ index.php?uf=23&dados=1

8. Cearense Foundation of Meteorology and Resources Hydric (1988) Bulletin climatic monitoring. FUNCEME, Fortaleza.

9. Ferreira, D.F. (2007) Sisvar version 5.0. System of statistical analysis. UFLA, Mines.

10. Chaves Filho, J.T. and Stacciarini-Seraphin, E. (2012) Alteration in osmotic potential and soluble carbohydrates content in young plants of lobeira (Solanumlycocarpum St.-Hil.) in response to water stress. http://www.scielo.br/scielo.php?script=sci_arttext&pid=S0100-84042001000200010&lng=en&nrm=iso

11. Fahn, A. and Cutler, D. (1992) Xerophytes. Gebruder Borntraeger, Berlin.

12. Pimentel, J.V.F. (2011) Growing cumaru (Amburanacearensis) in different production systems in the semiarid. Ph.D. Thesis, Federal University of Campina Grande, Campina Grande.

13. Silva, G. and Thieres, F., et al. (2011) Hydric demand and water use efficiency of sugar cane in the Brazilian semiarid irrigated. Rev. bras. eng. agric. ambient., Campina Grande.

14. Barros Jr., G., et al. (2008) Water consumption and use efficiency for two castor bean cultivars under hydric stress. Rev. bras. eng. agric. ambient., Campina Grande.

15. Lopes, O.D., et al. (2011) Determination of crop coefficient (Kc) and water use efficiency of irrigated rosemary pepper. Rev. bras. eng. agric. ambient., 15, 1415-4366.

# CHAPTER 9

# Somaclonal variations and their applications in horticultural crops improvement

**Hare Krishna[1], Mahdi Alizadeh[2], Dhurendra Singh[1], Udayvir Singh[1], Nitesh Chauhan[1], Maliheh Eftekhari[2], Radha Kishan Sadh[1]**

1 ICAR-Central Institute for Arid Horticulture, Beechwal, Bikaner, Rajasthan 334 006, India

2 Department of Horticulture, Faculty of Agriculture, Gorgan University of Agricultural Sciences and Natural Resources (GUASNR), Golestan, Gorgan, Iran

## ABSTRACT

The advancements made in tissue culture techniques has made it possible to regenerate various horticultural species in vitro as micropropagation protocols for commercial scale multiplication are available for a wide range of crops. Clonal propagation and preservation of elite genotypes, selected for their superior characteristics, require high degree of genetic uniformity amongst the regenerated plants. However, plant tissue culture may generate genetic variability, i.e., somaclonal variations as a result of gene mutation or changes in epigenetic marks. The occurrence of subtle somaclonal variation is a drawback for both in vitro cloning as well as germplasm preservation. Therefore, it is of immense significance to assure the genetic uniformity of in vitro raised plants at an early stage. Several strategies have been followed to ascertain the genetic fidelity of the in vitro raised progenies comprising morpho-physiological, biochemical, cytological and DNA-based molecular markers approaches. Somaclonal variation can pose a serious problem in any micropropagation program, where it is highly desirable to produce true-to-type plant material. On the other hand, somaclonal variation has provided a new and alternative tool to the breeders for obtaining genetic variability relatively rapidly and without sophisticated technology in horticultural crops, which are either difficult to breed or have narrow genetic base. In the present paper, sources of variations induced during tissue culture cycle and strategies to ascertain and confirm genetic fidelity in a variety of in vitro raised plantlets and potential application of variants in horticultural crop improvement are reviewed.

## KEYWORDS

Micropropagation; Somaclones; Oxidative stress; Epignetic variation; Molecular markers; Crop improvement

## 1. INTRODUCTION

Plant tissue culture techniques proffer a substitute method of vegetative propagation of horticultural crops (Krishna et al. 2005; Alizadeh et al. 2010). Clonal propagation through tissue culture (popularly known as micropropagation) can be realized relatively rapidly within a small space (Krishna et al. 2008; Eftekhari et al. 2012). The uniformity of individual plants within a clone population is a major advantage of clonal cultivars in commercial production (Krishna and Singh 2013). However, genetic variations do occur in undifferentiated cells, isolated protoplasts, calli, tissues and morphological traits of in vitro raised plants (Bairu et al. 2011; Currais et al. 2013). In 1981, Larkin and Scowkraft coined a general term "somaclonal variation" for plant variants derived from any form of cell or tissue cultures.

At present, micropropagated plants, in various crops, such as strawberry, papaya, banana, grapes, pineapple, citrus, tomato, cucumber, watermelon, rhododendron, orchids, etc., are preferred over plants propagated through conventional means. However, ever since the first formal report of morphological variants in sugarcane plants produced in vitro in 1971 (Heinze and Mee 1971), several instances of somaclonal variations have been reported in different horticultural crops. The notable example could be banana in which occurrence of off-types from tissue cultured plantlets ranged from 6 to 38 % in Cavendish cultivars (Sahijram et al. 2003); however, it could be as high as 90 % (Smith 1988). From the point of commercial micropropagation, variation of any kind, in particular, genetic variations may be considered obstructive and worthless; since, such variations may lead to loss of genetic fidelity. However, plant cell and tissue cultures render increased genetic variability comparatively faster and without applying a sophisticated technology. This technology holds ample scope in crop improvement of horticultural crops, which are largely propagated vegetatively, partly, due to reasons like longer juvenile phase as in perennial fruit crops, occasional inbreeding depression, self and cross incompatibility, narrow genetic base especially in ornamentals, etc. Further, somaclonal variations require less space and time for screening of desirable traits in vitro unlike cross seedlings of perennial crops, which require a great deal of land area and time. Somaclones may itself have numerous applications in plant breeding and genetic improvements and recovery of such novel variants can be enhanced by applying suitable in vitro selection pressure (Jain 2001; Lestari 2006).

## 2. SOURCES OF VARIATIONS DETECTED IN PLANT TISSUE CULTURE

Tissue culture is an efficient method of clonal propagation; however, the resulting regenerants often has a number of somaclonal variations (Larkin and Scowcroft 1981). These somaclonal variations are mainly caused by newly generated mutations arising from tissue culture process (Sato et al. 2011b). The triggers of mutations in tissue culture had been attributed to numerous stress factors, including wounding, exposure to sterilants during sterilization, tissue being incomplete (protoplasts as an extreme example), imbalances of media components such as high concentration of plant growth regulators (auxin and cytokinins), sugar from the nutrient medium as a replacement of photosynthesis in the leaves, lighting conditions, the disturbed relationship between high humidity and transpiration (Joyce et al. 2003; Sato et al. 2011b; Smulders and de Klerk2011).

Much of the variability expressed in micropropagated plants may be the result of, or related to, oxidative stress damage inflicted upon plant tissues during in vitro culture (Cassells and Curry 2001; Tanurdzic et al. 2008; Nivas and DSouza 2014). Oxidative stress results in elevated levels of pro-oxidants or reactive oxygen species (ROS) such as superoxide, hydrogen peroxide, hydroxyl, peroxyl and alkoxyl radicals. These ROS may involve in altered hyper- and hypo-methylation of DNA (Wacksman 1997); changes in chromosome number from polyploidy to aneuploidy, chromosome strand breakage, chromosome rearrangements, and DNA base deletions and substitutions (Czene and Harms-Ringdahl 1995), which in turn may lead to mutations in plant cells in vitro (Fig. 1). Somaclonal variation shows a similar spectrum of genetic variation to induced mutation as both of them result in qualitatively analogous gamut of DNA changes (Cassells et al. 1998). Different factors affect the frequency of development of somaclones under in vitro conditions.

## 2.1 Explant/explant source

Differences in both the frequency and nature of somaclonal variation may occur when regeneration is achieved from different tissue sources (Sahijram et al. 2003). Highly differentiated tissues such as roots, leaves, and stems generally produce more variations than explants with pre-existing meristems, such as axillary buds and shoot tips (Duncan1997). In general, the older and/or the more specialized the tissue is used for regeneration, the greater the chances that variation will be recovered in the regenerated plants (Table 1) as under such conditions, adventitious shoot regeneration (shoot organogenesis) takes place from atypical points of origin directly or indirectly through a callus stage (e.g., from leaves, petioles, shoot internodes, root segments, anthers, hypocotyls, cotyledons, etc.; Pijut et al. 2012). Somaclonal variation can also arise from somatic mutations already present in the donor plant, i.e., presence of chimera in explants (Karp 1994).

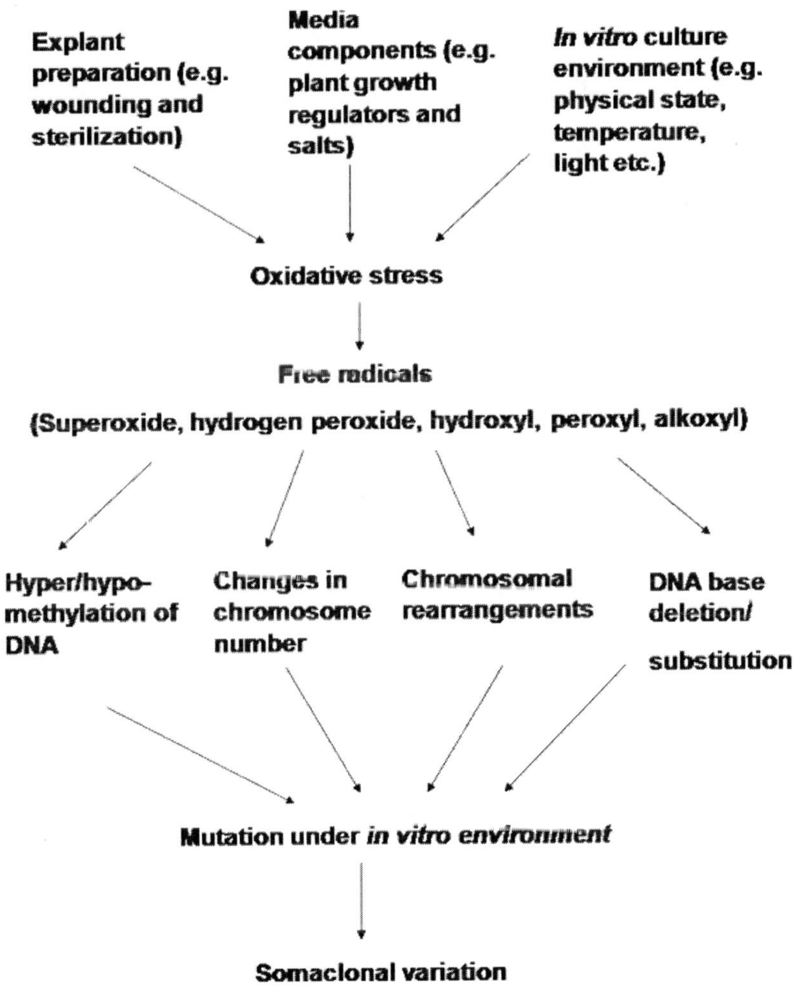


**Figure 1.** Mechanism of somaclonal variation in micropropagated plants as a result of oxidative burst upon in vitro culture

**Table 1.** Occurrence of somaclonal variations as affected by the choice of explants.

| S. no. | Crop species | Explants/explants source | Presence or absence of somaclonal variations (+/−) | References |
|---|---|---|---|---|
| 1 | African violet (*Saintpaulia* sp.) | Leaf segments | + | Matsuda et al. (2014) |
| 2 | Almond (*Prunus dulcis*) | Axillary branching | − | Martins et al. (2004) |
| 3 | Chimeric 'Maricongo' banana | Vegetative and floral axis tip | + | Krikorian et al. (1993) |
| | Cavendish group of bananas (*Musa* sp.) | Chimeric shoot tip | + | Israeli et al. (1995) |
| | Banana cv. Martaman | Shoot tip | − | Ray et al. (2006) |
| 4 | Brinjal (*Solanum melongena*) | Hypocotyl | − | Mallaya and Ravishankar (2013) |
| | | Callus induction on leaves, nodes and intermodal explants | + | Naseer and Mahmood (2014) |
| 5 | Chrysanthemum (*Dendranthema grandiflora*) | Callus from leaves and internodes | + | Miler and Zalewska (2014) |
| 6 | European violet (*Viola uliginosa* Besser) | Leaf and petiole fragments | + | Slazak et al. (2015) |
| 7 | Gerbera (*Gerbera jamesonii* Bolus) | Capitulum | − | Bhatia et al. (2009, 2011) |
| 8 | Gloxinia | Leaf explants | + | Hu and Xu (2010) |
| 9 | *Hedychium coronarium* Koen. | Axillary bud explants | − | Parida et al. (2013) |
| 10 | Hop (*Humulus lupulus* L.) | Meristem tissue | − | Patzak (2003) |
| 11 | *Kaempferia galanga* | Buds of rhizomes | − | Mohanty et al. (2011) |

| | | | | |
|---|---|---|---|---|
| 12 | Kiwifruit (*Actinidia deliciosa* (Chev.) Liang and Ferguson) cv. 'Tomuri' | Leaf blades and petioles | + | Prado et al. (2007) |
| 13 | Oil palm (*Elaeis guineensis* Jacq.) | Mature zygotic embryos | + | Rival et al. (2013) |
| | | Immature zygotic embryo | + | Sanputawong and Te-chato (2011) |
| | | Immature leaves | + | Lucia et al. (2011) |
| 14 | Papaya (*Carica papaya* L.) | Axillary shoot tips underwent cryopreservation | + | Kaity et al. (2009) |
| 15 | Patchouli (*Pogostemon patchouli*) | Callus induction on internodal and leaf explants | + | Ravindra et al. (2012) |
| 16 | Potato (*Solanum tuberosum*) | Callus cultures of stem explant | + | Thieme and Griess (2005) |
| | | Callus induction via fresh sprouts | + | Munir et al. (2011) |
| 17 | Sweet cherry (*Prunus avium*) | Shoot apical portions | + | Piagnani and Chiozzotto (2010) |
| 18 | Rootstock Mr.S 2/5, selected from a half-sib progeny from *Prunus cerasifera* Erhr | Leaf | + | Muleo et al. (2006) |
| 19 | *Swertia chirayita* | Axillary multiplication | – | Joshi and Dhawan (2007) |
| 20 | Turmeric (*Curcuma longa* L.) | Latent axillary buds of rhizome | – | Nayak et al. (2010) |
| | | Axillary buds of unsprouted rhizome | – | Panda et al. (2007) |
| | | Callus cultures established from rhizome segments | + | Kar et al. (2014) |
| 21 | *Vitis* spp. | Nodal segment | – | Alizadeh et al. (2008) |

## 3. MODE OF REGENERATION

Both culture initiation and subsequent subculture expose explants to oxidative stress (Krishna et al. 2008), which may result in mutations (Cassells and Curry 2001). It seems evident that 'extreme' procedures such as protoplast culture and also callus formation impose stress (Smulders and de Klerk 2011). Magnitude of this stress depends on the tissue culture technique. Therefore, the production of plants via axillary branching does not normally result in the production of variants, while cultures that go through a callus phase are the ones that theoretically promote a higher mutation rate (Zayova et al. 2010).

Investigations indicate more chromosome variability in the callus phase than in adventitious shoots (Saravanan et al. 2011), indicating a loss of competence in the more seriously disturbed genomes. This could be explained by the different grade of disturbance with which the cells are confronted. In the first case, cells follow a pattern of division which is the normal one in the developing plant. On the other hand, callus formation implies a dedifferentiation phase followed by uncontrolled cell divisions (Vázquez 2001). Some types of tissue culture mimic, in some aspects, other stressful situations as, for example, protoplast preparation in which cell wall degradation resembles the infective process of some pathogens. Therefore, the type and magnitude of the stress imposed on cultured cells varies according to the technique used. In contrast to popular belief that the growth of unorganized callus is necessary for induction of genetic variation, variability could be noticed in plants regenerated from explants adventitiously (Farahani et al. 2011; Bhojwani and Dantu 2013).

Sometimes for regeneration under in vitro conditions, somatic embryogenesis is the preferred pathway for generating propagules. It has been suggested that regeneration via embryogenesis has better chance of obtaining genetically uniform plants than through organogenic differentiation (Vázquez 2001). This is so, because DNA in the initial stages of development in somatic embryogenesis contains lower levels of methylation than in the later stages (Sahijram et al. 2003). Variation in in vitro cultures raised through somatic embryogenesis has been reported in several horticultural crops like hazel nut (Diaz-Sala et al. 1995), *Citrus paradisi* (Hao et al. 2004), oil palm (Jaligot et al. 2004), rose (Xu et al.2004), potato (Sharma et al. 2007), grapevine (Schellenbaum et al. 2008), coffee (Menéndez-Yuffá et al. 2010), olive (Leva et al. 2012), tamarillo (Currais et al. 2013) and brinjal (Naseer and Mahmood 2014).

# 4. EFFECT OF LENGTH OF CULTURE PERIOD AND NUMBER OF SUBCULTURE CYCLES

The longer a culture is maintained in vitro, the greater the somaclonal variation is (Kuznetsova et al. 2006; Gao et al. 2010; Farahani et al. 2011; Jevremović et al. 2012; Sun et al. 2013). Variant karyotypes are found to amass with increasing age of callus and as a result the chances of variant plants produced during successive subculture also increases, in general (Zayova et al. 2010). Furthermore, the rapid multiplication of a tissue, during micropropagation, may affect its genetic stability. Khan et al. (2011) reported that after the eighth subculture, the number of somaclonal variants increased with a simultaneous decrease in the multiplication rate of propagules in banana.

Similarly, Clarindo et al. (2012) suggested a limit of less than 4 months storage of coffee cell aggregate suspensions for true-to-type mass propagation as ploidy instability was noticed in long-term in vitro culture. Similarly when Farahani et al. (2011) raised olive cultivars, under in vitro conditions, through internode cuttings, significant difference was observed in morphological characters among the regenerated plants after seventh subculture, which was later confirmed by RAPD analysis. However, C-value analysis showed that no significant change has occurred during subculturing in both olive genotypes. This indicates that the genetic changes accompanied by somaclonal variation could be due to the changes in the nucleotide content of the genome, probably, owing to mutations (insertions/ deletions) and not due to quantitative changes.

Not only the number of subculture but their duration also contributes to enhancing the rate of somaclonal variations, especially cell suspension and callus cultures (Bairu et al.2006; Sun et al. 2013). Studies have shown that somaclonal variation is more apparent in plants regenerated from long-term cultures (Etienne and Bertrand 2003). Rival et al. (2013) noticed that in vitro proliferation induces DNA hypermethylation in a time-dependent fashion and changes in DNA methylation is involved in modulating the expression of embryogenic capacity of oil palm during tissue culture.

## 5. CULTURE ENVIRONMENT

External factors like growth regulators, temperature, light, osmolarity and agitation rate of the culture medium are known to influence the cell cycle in vivo in plants, considerably, which indicates that inadequate control of cell cycle in vitro is one of the causes of somaclonal variation (Karp 1994; Nwauzoma and Jaja 2013). Normal cell cycle controls, which prevent cell division before the completion of DNA replication, are presumed to be disrupted by tissue culture, resulting in chromosomal breakage (Phillips et al. 1994). Chromosome breakage and its consequences (deletions, duplications, inversions, and translocations) cause aberrations in vitro (Duncan 1997). Plant growth regulators can affect the rate of somaclonal variation both directly and indirectly by increasing the multiplication rate and inducing adventitious shoots (Gao et al. 2010). According to D'Amato (1985), it cannot be excluded that some plant growth regulators (PGRs) at certain concentrations or in combination with other growth regulators and/or particular constituents of a culture medium, may act as mutagens.

Several growth regulators, such as 2,4-dichlorophenoxy acetic acid (2,4-D), naphthalene acetic acid (NAA) and BAP (6-benzylaminopurine), synthetic phenylurea derivatives (4-CPPU, PBU and 2,3-MDPU) have been most frequently considered to be responsible for genetic variability (Siragusa et al. 2007; Sun et al. 2013; Sales and Butardo 2014).

Prolonged cultivation in medium containing 2,4-D influences higher DNA ploidy levels in callus cells (da Silva and Carvalho 2014). In their experiment with banana, Sales and Butardo (2014) observed that addition of synthetic auxin 2,4-D in culture medium led to high level of methylation events, particularly, cytosine methylation either at the internal or external cytosine end, which largely resulted in variations in tissue cultured plants. Alteration in genomic DNA methylation rate is being attributed for the development of 'mantled' somaclonal variant in oil palm (Eeuwens et al. 2002; Jaligot et al. 2011). Similarly, Arnhold-Schmitt (1993) observed that indole-3-acetic acid (IAA) and inositol in the growth medium induced DNA rearrangements and methylation changes in carrot (*Daucus carota*) callus cultures. Matsuda et al. (2014) observed that percentage of somaclonal variations dramatically increased when PGRs (0.5 ppm BA and 0.1 ppm NAA) were added to the medium inoculated with leaf/leaf segments explants of African violet.

Kinetin has been shown to cause extensive hypomethylation of DNA in proliferating cultures of carrot root explants within 2 weeks (Arnhold-Schmitt 1993), and auxins, including NAA, have the opposite effect and cause hypermethylation (LoSchiavo et al.1989). Moreover, there is evidence that differential expression in chromatin remodeling genes and histone methylation genes happens during tissue culture, which leads to disruption in the methylation pathway in a non-specific manner and hypo/hypermethylation patterns of DNA induced in tissue culture. This can be stabilized and transmitted to plants regenerated from these cultures (Shearman et al. 2013). Not only the concentration, but also the ratio of different growth regulators affects the occurrence of variations in vitro. Eeuwens et al. (2002) observed that, in general, a relatively high auxin/cytokinin ratio resulted in the lowest incidence of variant 'mantled' flowering in oil palm, while using media supplemented with relatively high cytokinins/auxin ratio resulted in a high incidence of mantled flowering. The role of cytokinin was further confirmed by Ooi et al. (2013), who noticed that the mantled inflorescences of oil palm contained higher levels of cytokinins like isopentenyladenine 9-glucoside and lower levels of trans-zeatin 9-glucoside, dihydrozeatin riboside, and dihydrozeatin riboside 5′-monophosphate compared with normal inflorescences.

# 6. GENOTYPE AND PLOIDY

Though, the in vitro morphogenesis seems to be highly dependent on plant growth regulators and media used for culture, it is again genotype specific (Alizadeh et al. 2010; Eftekhari et al. 2012). Among factors affecting somaclonal variation, plant genotype is probably the most important determinant of variation (Shen et al. 2007; Tican et al. 2008; Nwauzoma and Jaja 2013). Earlier, Eeuwens et al. (2002) characterized oil palm clones as low/moderate risk and high risk with regard to 'mantle' flowering (wherein anther primordia in both male and female flowers turn into fleshy supplementary carpels), on the basis of terminal inflorescence data generated under in vitro conditions. Clones classified as high risk at the outset gave a significantly higher incidence of mantled flowering in the field than low/medium risk clones, confirming that data on terminal inflorescences produced in vitro allows effective screening of material with regard to the risk of mantled flowering. It is likely that this result from a combination of differences in genotype and differences in epigenetically inherited changes are induced during the pre-embryogenic stages of the culture process, i.e., callus initiation and maintenance.

# 7. IDENTIFICATION OF VARIATION IN TISSUE CULTURE

Both genetic and epigenetic alterations are associated with in vitro propagation, which may have phenotypic consequences, and are collectively called somaclonal variation (Larkin and Scowcroft 1981; Guo et al. 2007). As a result, somaclonal variation is characterized by the intricacy of the changes, which are exhibited at various levels, including phenotypic, cytological, biochemical and genetic/epigenetic (Kaeppler et al.2000). Therefore, the strategy for the detection of somaclones should be based on such manifestations.

A wide variety of tools are available for the detection and characterization of somaclonal variants which are primarily based on the differences in morphological traits (Pérez et al.2009, 2011; Nhut et al.

2013), cytogenetical analysis for the determination of numerical and structural variation in the chromosomes (Clarindo et al. 2012; Currais et al. 2013; Abreu et al. 2014), biochemical (Vujovic et al. 2010; Kar et al. 2014), molecular DNA markers (Krishna and Singh 2007; Pathak and Dhawan 2012; Hossain et al. 2013; Bello-Bello et al. 2014) or their combinations (Horáček et al. 2013; Dey et al. 2015; Stanišić et al.2015). The best test for assessing somaclonal variation is to fruit out the plants and conduct an extensive horticultural evaluation, which is unfortunately a long-term endeavor with woody fruit crops, particularly (Grosser et al. 1996). Every tool has its own advantages and limitations in assessment of the variations (Table 2), which govern their use for restricted or large-scale application. The choice of technique for any given application depends upon the material used and the nature of the question being addressed (Karp 2000).

## 8. MOLECULAR BASIS OF SOMACLONAL VARIATION

How a single plant genotype can result in a variety of phenotypic outcomes under the same in vitro culture conditions is still far from being completely understood. Several bases for somaclonal variation have been proposed, which include changes in chromosome number (Mujib et al. 2007; Leva et al. 2012), point mutations (D'Amato 1985; Ngezahayo et al.2007), somatic crossing over and sister chromatid exchange (Duncan 1997; Bairu et al.2011), chromosome breakage and rearrangement (Czene and Harms-Ringdahl 1995; Alvarez et al. 2010), somatic gene rearrangement, DNA amplification (Karp 1995; Tiwari et al. 2013), changes in organelle DNA (Cassells and Curry 2001; Bartoszewski et al. 2007), DNA methylation (Guo et al. 2007; Linacero et al. 2011), epigenetic variation (Kaeppler et al. 2000; Guo et al. 2006; Smulders and de Klerk 2011), histone modifications and RNA interference (Miguel and Marum 2011), segregation of pre-existing chimeral tissue (Brar and Jain 1998; Vázquez 2001; Ravindra et al. 2012; Nwauzoma and Jaja 2013) and insertion or excision of transposable elements (Gupta 1998; Sato et al. 2011b). In particular, transposable elements are one of the causes of genetic rearrangements in in vitro culture (Hirochika et al. 1996; Sato et al. 2011a).

**Table 2.** Strengths and weaknesses of different marker systems for the assessment of clonal fidelity

| Advantages | Disadvantages |
|---|---|
| **Morphological traits** | |
| Visual differentiation | Sensitive to ontogenic changes and other environmental factors |
| Does not require any laboratory facility | Limited in numbers |
| Suitable for preliminary detection | Time-consuming |
| **Cytological markers (flow-cytometry)** | |
| Sample preparation and analysis is convenient and rapid in case of in flow-cytometry | Cytosolic compounds may interfere with quantitative DNA staining in flow-cytometry |
| Rapid and efficient method for routine large-scale studies of ploidy level | Absence of a set of internationally agreed DNA reference standards in case of in flow-cytometry |
| Unfailing detection of even the smallest modifications in chromosome number | Time-consuming chromosome counting |
| **Isozyme markers** | |
| Codominant expression | Sensitive to ontogenic changes and other environmental factors |
| Ease of performance | Limited in numbers |
| | Not all of these reagent systems work efficiently with all plant species |
| | Tissue-specific expression |
| **DNA markers** | |
| Codominant expression | RAPD markers are dominant and do not permit the scoring of heterozygous individuals. Besides, they exclusively identify sequence changes |
| Any source DNA can be used for the analysis | |
| Phenotypically neutral | Possible non-homology of similar sized fragments as ISSR is a multilocus technique |
| Not sensitive to ontogenic changes and other environmental factors | Disadvantages of AFLPs include the need for purified, high molecular weight DNA, the dominance of alleles and the possible non-homology of comigrating fragments belonging to different loci |
| Capability to detect culture-induced variation both at the DNA sequence and methylation pattern levels | Involvement of high development costs in SSR markers if adequate primer sequences for the crop species of interest are unavailable. Further, mutations in the primer annealing sites may result in the occurrence of null alleles (no amplification of the intended PCR product), which may lead to errors in scoring |

Tissue culture is reported to activate silent transposable elements, resulting in somaclonal variations. Insertions of transposable elements and retrotransposons can function as insertional mutagens of plant genomes, whereas widespread activation may result in a wide gamut of chromosomal rearrangements (Tanurdzic et al. 2008). In turn, these rearrangements can lead to misregulation of genes, aneuploidy and new transposon insertions (Smulders and de Klerk 2011).

However, many aspects of the mechanisms, which result in somaclonal variations, remain undefined. It is therefore, inevitable to explore the genome-wide change through sequencing of whole-genome of the concerned crop. Next-generation sequencing technology has enabled the whole-genome sequencing of individual plants (Miyao et al.2012). A new generation of sequencing technologies, from Illumina/Solexa, ABI/SOLiD, 454/Roche, and Helicos, has provided unprecedented opportunities for high-throughput functional genomic research (Morozova and Marra 2008; Metzker 2010).

# 9. SOMACLONAL VARIATIONS VIS-À-VIS CROP IMPROVEMENT

Genetic variation is an essential component of any conventional crop breeding program. The typical crop improvement cycle takes 10–15 years to complete and includes germplasm manipulations, genotype selection and stabilization, variety testing, variety increase, proprietary protection and crop production stages. Plant tissue culture is an enabling technology from which many novel tools have been developed to assist plant breeders (Karp 1992; Mathur 2013). Tissue culture-induced somaclonal variation is akin to variations induced with chemical and physical mutagens (Jain 2001) and offers an opportunity to uncover natural variability for their potential exploitation in crop improvement.

Like any other technology, in vitro induced somaclonal variation has its own merits and demerits, like the two sides of the same coin.

## 9.1 Advantages

The advantages comprise: (1) it is cheaper than other methods of genetic manipulation and does not require 'containment' procedures. (2) Tissue culture systems are available for more plant species than can be manipulated by somatic hybridization and transformation at the present time. (3) It is not necessary to have identified the genetic basis of the trait, or indeed, in the case of transformation, to have isolated and cloned it. (4) Novel variants have been reported among somaclones, and evidences indicate that both the frequency and distribution of genetic recombination events can be altered by passage though tissue culture. This implies that variation may be generated from different locations of the genome than those, which are accessible to conventional and mutation breeding (Karp1992). (5) There is no possibility of obtaining chimeric expression if somaclones are raised through cell culture (Evans 1989).

Somaclonal variation has been most successful in crops with limited genetic systems (e.g., apomicts, vegetative reproducers) and/or narrow genetic bases. In ornamental plants, for instance, the exploitation of in vitro-generated variability has become part of the routine breeding practice of many commercial enterprises.

## 9.2 Disadvantages

One of the serious limitations of somaclonal variation which makes it comparatively difficult to use is that, despite the identification of factors affecting the variation response of a given plant species, it is still not possible to predict the outcome of a somaclonal program (Karp 1992) as it is random and lacks reproducibility. Further, as a large number of genetic changes are based on point mutations or chromosome rearrangements, most $R_1$segregate. Therefore for quantitative traits such as yield, it is virtually impossible to select individuals with improvements in the $R_1$ generation. Though techniques for selection of somaclones resistant to various biotic and abiotic stresses had been worked out in many horticultural crops, unfortunately, no in vitro selection methods exist for complicated traits such as yield, soluble solids, sweetness, texture or shelf life (Evans 1989).

Somaclonal variation can become a part of plant breeding provided they are heritable and genetically stable. Only a limited numbers of promising varieties so far had been released using somaclonal variations. This is perhaps due to the lack of interaction between plant breeders and tissue culture scientists, and non-predictability of somaclones (Jain 2001). Further, though the new varieties have been produced by somaclonal variation, in a large number of cases improved variants have not been selected due to (1) the variations were all negative; (2) positive changes were also altered in negative ways; (3) the changes were not novel, or (4) the changes were not stable after selfing or crossing (Karp 1992).

# 10. RECOVERY OF SOMACLONAL VARIANTS

The recovery of variants can be improved by promoting the factors which are responsible for the development of somaclonal variations such as protoplast culture (Kothari et al.2010) and employing callus and cell suspension culture for several cycles and regeneration of large number of plants from long-term cultures (Barakat and El-Sammak 2011). Indirect organogenesis is an important means of retrieving genetic variation through somaclones with useful traits of agronomic or industrial use. Besides, plant genotype is a major factor, which determines the type and frequency of somaclonal variation. For instances, Solanaceous plants like potato (Sharma et al. 2007) and tomato (Bhatia et al.2005) produce a gamut of somaclonal variation than many other commercial horticultural crops. However, to be of practical value, the frequency of somaclonal variation should be sufficient enough to select desirable traits, and the selected lines should perform well under multiple environments (Duncan 1997). The efficiency of recovering variants in vitro can further be enhanced by applying selection pressure through screening of desirable traits, e.g., in vitro selection for tolerance against abiotic and biotic stresses (Barakat and El-Sammak 2011). This attains more significance in view of the fact that the selection of desirable traits takes several years and many generations under field conditions. In vitro selection can shorten considerably the time for the selection of desirable traits under in vitro selection pressure with minimal environmental interaction, and can complement field selection (Jain 2001).

The recovery of somaclones can be increased by combining micropropagation with induced mutagenesis in vitro (Afrasiab and Iqbal 2010). Kuksova et al. (1997) noted that somaclonal variation and mutagens can be combined to increase the frequency of induced mutation. Likewise, irradiation followed by adventitious bud regeneration has been reported to have allowed the recovery of mutants with useful agronomic traits in*Gypsophila paniculata* L. (Barakat and El-Sammak 2011). Yang and Schmidt (1994) treated in vitro leaves of the cherry rootstock '209/1' (*Prunus cerasus* × *P. canescens*) with X-rays with LD50 close to 20 Gy. Among plants regenerated from leaves with 20 Gy, one was phenotypically different, and was subsequently isolated and cloned. This somaclone was extremely dwarfed and was stable in both greenhouse and field tests. Employing more than one mutagen results in further improvement in recovery of somaclones in vitro. Murti et al. (2013) exposed the strawberry 'DNKW001' to the doses of 0, 30, 80, 130, 180, 230, 280, 300 and 325 Gy and similar doses of gamma rays + EMS 7 μM treatments. Their results showed that Gamma ray irradiation + EMS was more effective to generate more type and magnitude of variants. Purwati and Sudarsono (2007) regenerated four variant lines in abaca banana from (1) embryogenic calli; (2) ethyl methyl sulphonate (EMS)-treated embryogenic calli; (3) EMS-treated embryogenic calli, followed by in vitro selection on Foc (*Fusarium oxysporum* f.sp. cubense) culture filtrate (EMS + CF line) and (4) EMS-treated embryogenic calli, followed by in vitro selection on fusaric acid. The Foc resistance abaca variants were successfully identified from four tested abaca variant lines, although with different frequencies. However, more Foc resistance abaca plants were identified from EMS + CF line than the others. Earlier, Bidabadi et al. (2012) suggested that the subjecting of shoot tips cultures of banana to EMS (200 mM) treatments could provide an alternative strategy for inducing variants. Recently, Iuliana and Cerasela (2014) suggested irradiation of in vitro raised plants with ultraviolet radiations (UV-C) for induction of somaclones in potato.

# 11. APPLICATION OF SOMACLONAL VARIATIONS

It is well accepted that somaclonal variations arising out of unique tissue culture environment are very often noticed phenomenon in clonally propagated plants, which can advantageously be utilized as a source of new variation in horticultural crops (Karp 1995). However, suitable tools for detection, evaluation, identification and improvement of resistant clones should be designed in order to realize the benefits of such variations (Sahijram et al. 2003). Crop improvement through somaclonal variation enables breeders to obtain plants tolerant to the biotic or abiotic stress, such as drought, high salinity, high or low soil pH and disease tolerance (Yusnita et al. 2005). A number of cultivars have been developed through somaclonal variation in different horticultural crops for a range of useful traits, which are presented in Table 3.

# 12. CONCLUSIONS

Several strategies have been followed to ascertain the genetic fidelity of the in vitro produced progenies in view of the fact that the commercial viability of micropropagation technology is reliant upon maintenance of genetic fidelity in the regenerated plants. Therefore, a thorough assessment of micropropagated plants becomes very critical, especially, for perennial crops such as fruit species, which have a long pre-bearing growth period. The efficiency and sensitivity of new molecular tools has enabled us to detect somaclonal variation at an early stage. These tools have become very useful for the rapid detection and accurate identification of variants. Nevertheless, the morphological and cytological assays should continue to remain as the primary and essential assay for the sustained success of fidelity tests associated with production of clonal plants. Though, on one hand, tissue culture-induced variations pose a major threat to the genomic integrity of regenerated plants, they provide tools for improvement to plant breeders, particularly for crops with a narrow genetic base, i.e., self pollinated and vegetatively propagated. Irrespective of our goal either for production of true-to-the type planting material or creation of variability, a multidisciplinary approach (involving concerned sciences of horticulture, genetics and plant breeding, physiology, cytology and molecular biology) with all our previous knowledge and experience should be followed to achieve the desideratum.

**Table 3.** In vitro selection of desirable traits and development of some commercially exploited varieties through somaclonal variation in different horticultural crops

| S. no. | Horticultural crop | Characteristic of somaclone | References |
|---|---|---|---|
| 1 | *Aglaonema* | Cultivar 'Moonlight Bay' and 'Diamond Bay' from 'Silver Bay,' and 'Emerald Bay,' from 'Golden Bay' | Henny et al. (1992, 2003) |
| 2 | Apple (*Malus × domestica* Borkh.) | Resistance to *Erwinia amylovora* | Chevreau et al. (1998) |
| 3 | Apple rootstocks M 26 and MM 106 (*Malus pumila* Mill.) | Resistance to *Phytophthora cactorum* | Rosati et al. (1990) |
| 4 | Apple rootstock Malling 7 | Resistance to white root rot (*Dematophora necatrix*) | Modgil et al. (2012) |
| 5 | *Anthurium* sp. | 'Orange Hot' derived from 'Red Hot' clone | Henny and Chen (2011) |
| 6 | Banana (*Musa acuminata* L.) | Semi-dwarf and resistant to *Fusarium* wilt TC1-229 | Tang et al. (2000) |
| | | Larger bunch size var. TC2-425; Resistant to *Fusarium oxysporum* f. sp. cubense (Foc) race 4; bunch 40 % heavier than cv. Formosana | Hwang (2002) |
| | | *Fusarium* wilt-resistant somaclonal variants of banana cv. Rasthali | Ghag et al. (2014) |
| | | Var. CIEN-BTA-03, resistant to yellow Sigatoka | Giménez et al. (2001) |
| | | 10 somaclones: GCTCV215-1 released for commercial planting | Hwang and Ko (1992, 2004) |
| | | Var. CUDBT-B1, reduced height and early flowering | Martin et al. (2006) |
| | | Var. Tai-Chiao No. 5, superior horticultural traits and resistance to *Fusarium* wilt | Lee et al. (2011) |
| 7 | Begonia (*Begonia × elatior*) | Plant morphology, number of flowers per plant, and flower size | Jain (1997) |
| 8 | Brinjal (*Solanum melongena* L.) | Stress-tolerant somaclone selection | Ferdausi et al. (2009) |
| 9 | Blackberry | Thornless var. 'Lincoln Logan' | Hall et al. (1986) |
| 10 | Capsicum (*Capsicum annuam* L.) | Yellow fruited var. Bell sweet | Morrison et al. (1989) |
| 11 | *Calthea roseopicta* | Developed common cultivars like Angela, Cora, Dottie, Eclipse and Saturn | Chao et al. (2005) |
| 12 | Carrot (*Daucus carota* L.) | Resistance to leaf spot (*Alternaria dauci*) | Dugdale et al. (2000) |
| | | Resistant to drought | Rabiei et al. (2011) |
| 13 | Carnation (*Dianthus caryoplyllus* L.) | Resistant to *Fusarium oxysporum* f. sp. *dianthi* | Esmaiel et al. (2012) |
| 14 | Celery (*Apium graveolens* L.) | *Fusarium* resistant var. UC-TC | Heath-Pagliuso and Rappaport (1990) |
| | | Multiple-resistant (insect resistance against *Spodoptera exigua* and disease resistance against *Fusarium* yellow) somaclones K-26, K-108 and K-128 | Diawara et al. (1996) |
| 15 | *Celosia argentea* L. | Resistance to nematode | Opabode and Adebooye (2005) |
| 16 | *Cereus peruvianus* | Shoots with different areoles characteristics | Resende et al. (2010) |
| 17 | Chili pepper (*Capsicum annuum* L.) | Early flowering and increase of yield components | Hossain et al. (2003) |
| 18 | Chrysanthemum (*Dendranthema grandiflora*) | Variation in leaf, flower shape and petal size | Ahloowalia (1992) |
| | | Daisy type chrysanthemum | Jevremović et al. (2012) |
| | | Attractive variants with changed inflorescence colors | Miler and Zalewska (2014) |
| 19 | *Citrus* spp. | Resistant to *Phoma tracheiphila* | Deng et al. (1995) |
| | | Salinity tolerance | Ben-Hayyim and Goffer (1989) |
| 20 | *Cuphea viscosissima* Jacq. | Significantly superior over the parents for mean plant height, leaf area, seed yield, per cent caprylic acid and lauric acid contents | Ben-Salah and Roath (1994) |

| | | | |
|---|---|---|---|
| 21 | *Cymbopogon winterianus* Jowitt | Aromatic grass var. CIMAP/Bio-13 with 50–60 % increased oil yield | Mathur et al. (1988) |
| | | Increased total oil yield and quality with high geraniol content | Nayak et al. (2003) |
| | *Cymbopogon martinii* | Increased oil content | Patnaik et al. (1999) |
| 22 | *Dieffenbachia* sp. | Novel and distinct foliar variegation with taller, larger canopy and longer leaves than 'Camouflage' parental plants | Shen et al. (2007) |
| 23 | Garlic (*Allium sativum* L.) | Consistently higher bulb yield than the parental clone | Vidal et al. (1993) |
| | | Resistance against the pathogenic fungi '*Sclerotium cepivorum*' | Zhang et al. (2012) |
| 24 | *Geranium* spp. | Vigourous and attractive flower | Skirvin and Janick (1976) |
| | | Isomenthone-rich somaclonal mutant | Gupta et al. (2001) |
| | | Cv. 'CIM Pawan, a somaclone of the Bourbon type variety Bipuli, with more herbage and essential oil yield than Bipuli | Saxena et al. (2008) |
| 25 | Gerbera (*Gerbera jamesonii* Bolus) | Novel cultivars | Minerva and Kumar (2013) |
| 26 | Ginger (*Zingiber officinale* Rosc.) | Tolerant to wilt pathogen (*Fusarium oxysporum* f.sp. *zingiberi* Trujillo) | Bhardwaj et al. (2012) |
| 27 | Grapevine (*Vitis vinifera* L.) | Resistant to *Botrytis cinerea* and *Plasmopara viticola* | Kuksova et al. (1997) |
| 28 | *Haemerocallis* spp. | Dwarf, short flowers, male sterile var. Yellow Tinkerbell | Griesbach (1989) |
| 29 | *Hedychium* (ornamental ginger) | Ramata, dwarf and variegated cultivar | Sakhanokho et al. (2012) |
| 30 | *Java citronella* (*Cymbopogon winterianus*) | Somaclonal variant variety CIMAP/Bio-13, which yields 37 % more oil and 39 % more citronellon than the control variant | Mathur (2010) |
| 31 | Kiwi fruit (*Actinidia deliciosa*) | 5 somaclones, derived from cv. Tamuri, tolerant to NaCl | Caboni et al. (2003) |
| 32 | Mango (*Mangifera indica* L.) | Resistant to *Colletotrichum gleosporiensis* | Litz et al. (1991) |
| 33 | Mint (*Mentha arvensis*) | Increased herb and oil yield | Kukreja et al. (1991; 2000) |
| 34 | Myrobolan (*Prunus cerasifera* Erhr) | Water logging-tolerant clone variant (S.4) of myrobolan rootstock Mr.S 2/5 for peach cv. Sun Crest | Iacona et al. (2013) |
| 35 | Olive (*Olive europea*) | Bush olive somaclone (BOS), columnar olive somaclone (COS) | Leva et al. (2012) |
| 36 | Patchouli (*Pogostemon patchouli*) | Higher herb yield and essential oil content | Ravindra et al. (2012) |
| 37 | Pea (*Pisum sativum* L.) | Resistance to *Fusarium solani* | Horáček et al. (2013) |
| 38 | Peach (*Prunus persica* L.) | Somaclones E1F6 and E122 resistant to leaf spot, moderately resistant to canker in cvs. Sunhigh and Red haven | Hammerschlag and Ognjanov (1990) |
| | | Resistant to root-knot nematode (*Meloidogyne incognita* Kofoid and White) | Hashmi et al. (1995) |
| | | Somaclone S 122-1 was found resistant to bacterial canker (*Pseudomonas syringae* pv. *syringae*) | Hammerschlag (2000) |
| 39 | Pear (*Pyrus* sp.) | Resistant to *Erwinia amylovora* | Viseur (1990) |
| | Pear rootstock (*Pyrus communis* L.) 'Old Home × Farmingdale (OHF 333)' | Tolerance to the fire blight | Nacheva et al. (2014) |
| 40 | *Philodendron* | Cultivars 'Baby Hope' from 'Hope' | Devanand et al. (2004) |
| 41 | *Picrorhiza kurroa* | Higher glycoside contents including kutkoside and picroside I in somaclone 14-P derived through *Agrobacterium rhizogenes* mediated transformed hairy root cultures of *P. kurroa* | Mondal et al. (2013) |

| | | | |
|---|---|---|---|
| 42 | Pineapple (*Ananas comosus* L., Merr.) | Spineless variant | Jaya et al. (2002) |
| | | Cvs. P3R5 and Dwarf, variation in fruit color, growth habit, fruit size and length of plant generation cycle | Pérez et al. (2009, 2012) |
| 43 | Potato (*Solanum tuberosum* L.) | Non-browning var. White Baron | Arihara et al. (1995) |
| | | Somaclones for heat tolerance | Das et al. (2000) |
| | | Somaclones IBP-10, IBP-27 and IBP-30, derived from cultivar Desiree, showed higher resistance to *Alternaria solani* and *Streptomyces scabiei* | Veitia-Rodriguez et al. (2002) |
| | | Improved size, shape, appearance, starch content and starch yield | Thieme and Griess (2005) |
| | | Superior processing attributes than cv. 'Russet Burbank' | Nassar et al. (2011) |
| | | High-yielding genotype SVP-53 | Hoque and Morshad (2014) |
| | | Increased phytonutrient and antioxidant components over cv. 'Russet Burbank' | Nassar et al. (2014) |
| 44 | Quince A (*Cydonia oblonga*) | High soil pH | Dolcet-Sanjuan et al. (1992), Marino et al. (2000) |
| 45 | *Stevia rebaudiana* | High glycoside contents (steviol, stevioside, and rebaudioside) | Khan et al. (2014) |
| 46 | Strawberry (*Fragaria* sp.) | Resistant to *Fusarium oxysporum f. sp. fragariae* | Toyoda et al. (1991) |
| | | Resistant to *Alternaria alternate* | Takahashi et al. (1993) |
| | | Resistant to *Phytophthora cactorum* | Battistini and Rosati (1991) |
| | | Improved horticultural traits | Biswas et al. (2009) |
| | | Resistant to *Verticillium dahliae* Kleb | Zebrowska (2010) |
| | | 'Serenity', a paler skin-colored, late season, resistant to powdery mildew and *Verticillium* wilt somaclonal variant of the short-day cv. 'Florence' | Whitehouse et al. (2014) |
| 47 | Sweet potato (*Ipomea batatas* L. Lam.) | Tolerant to salinity | Anwar et al. (2010) |
| 48 | Sweet orange (*Citrus sinensis* (L.) Osb.) | Somaclone of OLL (Orie Lee Late) sweet orange; late maturing; suitable for fresh market or processing, exceptional juice quality and flavor | Grosser et al. (2015) |
| 49 | St. Augustine grass [*Stenotaphrum secundatum* (Walt.) Kuntze] | Freeze-tolerant somaclonal variant SVC3 | Li et al. (2010) |
| 50 | *Syngonium podophyllum* Schott | 22 cultivars, derived from original 'White Butterfly' clone, with distinct and stable foliage characteristics | Henny and Chen (2011) |
| 51 | Tomato (*Lycopersicon esculentum* L.) | High solid contents var. DNAP9 | Evans (1989) |
| 52 | Tulip (*Tulipa* sp.) | "Bs6", selected from among the micropropagated plants of the cultivar 'Blue Parrot' with red-violet colored longer flower and stem | Podwyszynska et al. (2010) |
| 53 | Torenia (*Torenia fournieri*) | Flower color somaclonal variants | Nhut et al. (2013) |
| 54 | Turmeric (*Curcuma longa* L.) | High essential oil yielding somaclones | Kar et al. (2014) |
| | | Turmeric somaclone resistant to *Fusarium oxysporum* f.sp. Zingiberi | Kuanar et al. (2014) |
| 55 | Indian ginseng (*Withania somnifera* (L.) Dunal) | Withanolide (12-deoxywithastramonolide)-rich somaclonal variant | Rana et al. (2012) |


# ACKNOWLEDGMENTS

Authors are grateful to the Dr. S.K. Singh, Principal Scientist, Division of Fruits and Horticultural Technology, Indian Agricultural Research Institute, New Delhi 12, India, for his valuable advice during the preparation of the manuscript.

# REFERENCES


1. Abreu IS, Carvalho CR, Clarindo WR (2014) Massal induction of *Carica papaya* L. 'Golden' somatic embryos and somaclone screening by flow cytometry and cytogenetic analysis. Cytologia 79(4):475–484CrossRef

2. Afrasiab H, Iqbal J (2010) In vitro techniques and mutagenesis for the genetic improvement of potato cvs. Desiree and Diamant. Pak J Bot 42:1629–1637

3. Ahloowalia BS (1992) *In vitro* radiation induced mutants in *Chrysanthemum*. Mutat Breed Newsl 39:6

4. Alizadeh M, Singh SK, Jhang T, Sharma TR (2008) Inter simple sequence repeat analysis to confirm genetic stability of micropropagated plantlets in three grape (*Vitis* spp) rootstock genotypes. J Plant Biochem Biotechnol 17:77–80

5. Alizadeh M, Singh SK, Patel VB (2010) Comparative performance of in vitro multiplication in four grape (*Vitis* spp.) rootstock genotypes. Int J Plant Prod 4:41–50

6. Alvarez ME, Nota F, Cambiagno DA (2010) Epigenetic control of plant immunity. Mol Plant Pathol 11:563–576

7. Anwar A, Kikuchi A, Watanabe KN (2010) Assessment of somaclonal variation for salinity tolerance in sweet potato regenerated plants. Afr J Biotechnol 9:7256–7265

8. Arihara A, Kita T, Igarashi S, Goto M, Irikura Y (1995) White Baron: a non-browning somaclonal variant of Danshakuimo (Irish cobbler). Am Potato J 72(11):701–705

9. Arnhold-Schmitt B (1993) Rapid changes in amplification and methylation pattern of genomic DNA in cultured carrot root explants (*Daucus carota* L.). Theor Appl Genet 85:793–800

10. Bairu MW, Fennell CW, van Staden J (2006) The effect of plant growth regulators on somaclonal variation in Cavendish banana (*Musa* AAA cv. 'Zelig'). Sci Hortic 108:347–351

11. Bairu MW, Aremu AO, Staden JV (2011) Somaclonal variation in plants: causes and detection methods. Plant Growth Regul 63:147–173

12. Barakat MN, El-Sammak H (2011) In vitro mutagenesis, plant regeneration and characterization of mutants via RAPD analysis in baby's breath '*Gypsophila paniculata* L'. Aust J Crop Sci 5(2):214–222

13. Bartoszewski G, Havey MJ, Ziókowska A, D'ugosz M, Malepszy S (2007) The selection of mosaic (MSC) phenotype after passage of cucumber (*Cucumis sativus* L.) through cell culture—a method to obtain plant mitochondrial mutants. J Appl Genet 48:1–9

14. Battistini C, Rosati P (1991) *In vitro* evaluation of somaclonal strawberry (*Fragaria* × *ananassa* 'Brighton') variants for susceptibility to *Phytophthora cactorum*. In: Dale A, Lubby D (eds) The strawberry into the 21st century. Timber Press, Portland, pp 121–123

15. Bello-Bello JJ, Iglesias-Andreu LG, Avilés-Viñas SA, Gómez-Uc E, Canto-Flick A, Santana-Buzzy N (2014) Somaclonal variation in habanero pepper (*Capsicum chinense* Jacq.) as assessed ISSR molecular markers. HortScience 49(4):481–485

16. Ben-Hayyim G, Goffer Y (1989) Plantlet regeneration from NaCl-selected salt-tolerant callus culture of Shamouti oranges (*Citrus sinesis* L. Osbeck). Plant Cell Rep 7:680–683

17. Ben-Salah H, Roath WW (1994) Somaclonal variation in *Cuphea viscosissima* Jacq. for plant improvement. Ind Crop Prod 2:239–244

18. Bhardwaj SV, Thakur T, Sharma R, Sharma P (2012) In vitro selection of resistant mutants of ginger (*Zingiber officinale* Rosc.) against wilt pathogen (*Fusarium oxysporum* f. sp.*zingiberi* Trujillo). Plant Dis Res 27:194–199

19. Bhatia P, Ashwath N, Senaratna T, Krauss SL (2005) Genetic analysis of cotyledon derived regenerants of tomato using AFLP markers. Curr Sci 88(2):280–284

20. Bhatia R, Singh KP, Jhang T, Sharma TR (2009) Assessment of clonal fidelity of micropropagated gerbera plants by ISSR markers. Sci Hortic 119:208–211

21. Bhatia R, Singh KP, Sharma TR, Jhang T (2011) Evaluation of the genetic fidelity of in vitro-propagated gerbera (*Gerbera jamesonii* Bolus) using DNA-based markers. Plant Cell Tissue Organ Cult 104(1):131–135

22. Bhojwani SS, Dantu PK (2013) Plant tissue culture: an introductory text. Springer, India

23. Bidabadi SS, Meon S, Wahab Z, Subramaniam S, Mahmood M (2012) Induced mutations for enhancing variability of banana (*Musa* spp.) shoot tip cultures using ethyl methanesulphonate (EMS). Aust J Crop Sci 6:391–401

24. Biswas MK, Dutt M, Roy UK, Islam R, Hossain M (2009) Development and evaluation of in vitro somaclonal variation in strawberry for improved horticultural traits. Sci Hortic 122:409–416

25. Brar DS, Jain SM (1998) Somaclonal variation: mechanism and applications in crop improvement. In: Jain SM, Brar DS, Ahloowalia BS (eds) Somaclonal variation and induced mutations in crop improvement. Kluwer Academic Publishers, Dordrecht, pp 15–37

26. Caboni E, Anselmi S, Donato E, Manes F (2003) In vitro selection of *Actinidia deliciosa*clones tolerance to NaCl and their molecular and in vivo ecophysiological characterisation. Acta Hortic 618:77–83

27. Cassells AC, Curry RF (2001) Oxidative stress and physiological, epigenetic and genetic variability in plant tissue culture: implications for micropropagators and genetic engineers. Plant Cell Tissue Organ Cult 64:145–157

28. Cassells AC, Deadman ML, Brown CA, Griffin E (1998) Field resistance to late blight (*Pytophtora infectans* (Mont.) De Bary in potato (*Solanum tuberosum* L.) somaclones associated with instability and pleiotropic effects. Euphytica 57:157–167

29. Chao CCT, Devanand PS, Chen J (2005) AFLP analysis of genetic relationships among*Calathea* species and cultivars. Plant Sci 168:1459–1469

30. Chevreau E, Brisset MN, Paulin JP, James DJ (1998) Fire blight resistance and genetic trueness-to-type of four somaclonal variants from the apple cultivar Greensleeves. Euphytica 104:199–205

31. Clarindo WR, Carvalho CR, Mendonça MAC (2012) Ploidy instability in long-term in vitro cultures of *Coffea arabica* L. monitored by flow cytometry. Plant Growth Regul 68(3):533–538

32. Currais L, Loureiro J, Santos C, Canhoto JM (2013) Ploidy stability in embryogenic cultures and regenerated plantlets of tamarillo. Plant Cell Tissue Organ Cult 114:149–159

33. Czene M, Harms-Ringdahl M (1995) Detection of single-strand breaks and formamidoprymidine-DNA glycosylase-sensitive sites in DNA of cultured human fibroblasts. Mutat Res 336:235–242

34. D'Amato F (1985) Cytogenetics of plant cell and tissue cultures and their regenerants. CRC Crit Rev Plant Sci 3:73–112

35. da Silva TCR, Carvalho CR (2014) Vertical heterogeneity of DNA ploidy level assessed by flow cytometry in calli of *Passiflora cincinnata*. In Vitro Cell Dev Biol Plant 50(2):158–165

36. Das A, Gosal SS, Sidhu JS, Dhaliwal HS (2000) Induction of mutations for heat tolerance in potato by using in vitro culture and radiation. Euphytica 114:205–209

37. Deng ZN, Gentile A, Domina F, Nicolosi E, Tribulato E (1995) Selecting lemon protoplasts for insensitivity to *Phoma tracheiphila* toxin and regenerating tolerant plants. J Am Soc Hortic Sci 120:902–905

38. Devanand PS, Chen J, Henny RJ, Chao CCT (2004) Assessment of genetic relationships among *Philodendron* cultivars using AFLP markers. J Am Soc Hortic Sci 129:690–697

39. Dey T, Saha S, Ghosh PD (2015) Somaclonal variation among somatic embryo derived plants—evaluation of agronomically important somaclones and detection of genetic changes by RAPD in *Cymbopogon winterianus*. S Afr J Bot 96:112–121

40. Diawara MM, Trumble JT, Lacy ML, White KK, Carson WG (1996) Potential of somaclonal celeries for use in integrated pest management. J Econ Entomol 89:218–223

41. Diaz-Sala C, Rey M, Boronat A, Besford R, Rodriguez R (1995) Variations in the DNA methylation and polypeptide patterns of adult hazel (*Corylus avellana* L.) associated with sequential in vitro subcultures. Plant Cell Rep 15:218–221

42. Dolcet-Sanjuan R, Mok DWS, Mok MC (1992) Characterization and in vitro selection for iron efficiency in *Pyrus* and *Cydonia*. In Vitro Cell Dev Biol Plant 28:25–29

43. Dugdale LJ, Mortimer AM, Isaac S, Collin HA (2000) Disease response of carrot and carrot somaclones to *Alternaria dauci*. Plant Pathol 49:57–67

44. Duncan RR (1997) Tissue culture-induced variation and crop improvement. Adv Agron 58:201–240

45. Eeuwens CJ, Lord S, Donough CR, Rao V, Vallejo G, Nelson S (2002) Effects of tissue culture conditions during embryoid multiplication on the incidence of "mantled" flowering in clonally propagated oil palm. Plant Cell Tissue Organ Cult 70:311–323

46. Eftekhari M, Alizadeh M, Mashayekhi K, Asghari HR (2012) In vitro propagation of four Iranian grape varieties: influence of genotype and pretreatment with arbuscular mycorrhiza. Vitis 51:175–182

47. Esmaiel NM, Al-Doss AA, Barakat MN (2012) In vitro selection for resistance to *Fusarium oxysporum* f. sp. *dianthi* and detection of genetic polymorphism via RAPD analysis in carnation. J Med Plants Res 6(23):3997–4004

48. Etienne H, Bertrand B (2003) Somaclonal variation in *Coffea arabica*: effects of genotype and embryogenic cell suspension age on frequency and phenotype of variants. Tree Physiol 23:419–426

49. Evans DA (1989) Somaclonal variation—genetic basis and breeding applications. Trends Genet 5:46–50

50. Farahani F, Yari R, Masoud S (2011) Somaclonal variation in Dezful cultivar of olive (*Oleaeuropaea* subsp. *europaea*). Gene Conserve 10:216–221

51. Ferdausi A, Nath UK, Das BL, Alam MS (2009) In vitro regeneration system in brinjal (*Solanum melongena* L.) for stress tolerant somaclone selection. J Bangladesh Agric Univ 7(2):253–258

52. Gao X, Yang D, Cao D, Ao M, Sui X, Wang Q, Kimatu JN, Wang L (2010) In vitro micropropagation of *Freesia hybrid* and the assessment of genetic and epigenetic stability in regenerated plantlets. J Plant Growth Regul 29:257–267

53. Ghag SB, Shekhawat UK, Ganapathi TR (2014) Characterization of Fusarium wilt resistant somaclonal variants of banana cv. *Rasthali* by cDNA-RAPD. Mol Biol Rep 41(12):7929–7935

54. Giménez C, Garcia DE, Enrech NX, Blanca I (2001) Somaclonal variation in banana: cytogenetic and molecular characterization of the somaclonal variant CIEN BTA-03. In Vitro Cell Dev Biol Plant 37:217–222

55. Griesbach RJ (1989) Selection of a dwarf Hemerocallis through tissue culture. HortScience 24:1027–1028

56. Grosser JW, Gmitter FG Jr, Chandler JL (1996) Somaclonal variation in sweet orange—a tool for cultivar improvement. In:

Proceedings of International Society of Citriculture VIII Congress, Sun City, South Africa, pp 203–206

57. Grosser JW, Gmitter FG Jr, Dutt M, Calovic M, Ling P, Castle B (2015) Highlights of the University of Florida, Citrus Research and Education Center's comprehensive citrus breeding and genetics program. Acta Hortic 1065:405–413

58. Guo W, Gong L, Ding Z, Li Y, Li F, Zhao S, Liu B (2006) Genomic instability in phenotypically normal regenerants of medicinal plant *Codonopsis lanceolata* Benth. et Hook. f., as revealed by ISSR and RAPD markers. Plant Cell Rep 25:896–906

59. Guo W, Wu R, Zhang Y, Liu X, Wang H, Gong L, Zhang Z, Liu B (2007) Tissue culture-induced locus-specific alteration in DNA methylation and its correlation with genetic variation in *Codonopsis lanceolata* Benth. et Hook. f. Plant Cell Rep 26:1297–1307

60. Gupta PK (1998) Chromosomal basis of somaclonal variation in plants. In: Jain SM, Brar DS, Ahloowalia BS (eds) Somaclonal variation and induced mutations in crop improvement. Kluwer Academic Publishers, Dordrecht, pp 149–168

61. Gupta R, Mallavarapu GR, Banerjee S, Kumar S (2001) Characteristics of an isomenthone-rich somaclonal mutant isolated in a geraniol-rich rose-scented geranium accession of*Pelargonium graveolens*. Flavour Frag J 16(5):319–324

62. Hall HK, Skirvin RM, Braam WF (1986) Germplasm release of 'Lincoln Logan', a tissue culture-derived genetic thornless 'Loganberry'. Fruit Var J 40:134–135

63. Hammerschlag FA (2000) Resistant responses of peach somaclone 122–1 to *Xanthomonas campestris* pv. *pruni* and to *Pseudomonas syringae* pv. *syringae*. HortScience 35:141–143

64. Hammerschlag FA, Ognjanov V (1990) Somaclonal variation in peach: screening for resistance to *Xanthomonas camperstris* pv. *pruni* and *Pseudomonas syringae* pv.*syringae*. Acta Hortic 280:403–408

65. Hao YJ, Wen XP, Deng XX (2004) Genetic and epigenetic evaluations of citrus calluses recovered from slow-growth culture. J Plant Physiol 161:479–484

66. Hashmi G, Hammerschlag FA, Huettel RN, Krusberg LR (1995) Growth, development and response of peach somaclones to the root-knot nematode, *Meloidogyne incognita.* J Am Soc Hortic Sci 120:932–937

67. Heath-Pagliuso S, Rappaport L (1990) Somaclonal variant UC-T3: the expression of*Fusarium* wilt resistance in progeny arrays of celery, *Apium graveolens* L. Theor Appl Genet 80:390–394

68. Heinze DJ, Mee GWP (1971) Morphologic, cytogenetic and enzymatic variation in*Saccharum* species hybrid clones derived from callus culture. Am J Bot 58:257–262

69. Henny RJ, Chen J (2011) Tropical foliage plant development: origin of new cultivars. ENH1092. Environmental Horticulture Department, Florida Cooperative Extension Service, Institute of Food and Agricultural Sciences, University of Florida.http://edis.ifas.ufl.edu

70. Henny RJ, Poole RT, Conover CA (1992) 'Silver Bay' *Aglaonema.* HortScience 27:1238

71. Henny RJ, Chen J, Norman DJ (2003) 'Diamond Bay' and 'Emerald Bay'*Aglaonema.* HortScience 38:1446–1447

72. Hirochika H, Sugimoto K, Otsuki Y, Tsugawa H, Kanda M (1996) Retrotransposons of rice involved in mutations induced by tissue culture. Proc Natl Acad Sci USA 93:7783–7788

73. Hoque ME, Morshad MN (2014) Somaclonal variation in potato (*Solanum tuberosum* L.) using chemical mutagens. Agriculturists 12(1):15–25

74. Horáček J, Švábová L, Šarhanová P, Lebeda A (2013) Variability for resistance to*Fusarium solani* culture filtrate and fusaric acid among somaclones in pea. Biol Plant 57(1):133–138

75. Hossain MA, Konisho K, Minami M, Nemoto K (2003) Somaclonal variation of regenerated plants in chili pepper (*Capsicum annuum* L.). Euphytica 130:233–239

76. Hossain MM, Kant R, Van PT, Winarto B, Zeng S, Teixeira da Silva JA (2013) The application of biotechnology to orchids. Crit Rev Plant Sci 32(2):69–139

77. Hu X, Xu Q (2010) Morphological Variations of somaclone regenerated from series tissue culture of *Gloxinia*. Acta Agric Boreal Occident Sin 5:39

78. Hwang S (2002) Somaclonal variational approach to breeding Cavendish banana for resistance to *Fusarium* wilt race 4. In: Global conference on banana and plantain, Bangalore, India, Abstract 57, October 28–31

79. Hwang SC, Ko WH (1992) Breeding for resistance to *Fusarium* wilt of Cavendish banana by using tissue culture method. Spec Publ Taichung Dist Agric Improv Stn 29:229–237

80. Hwang SC, Ko WH (2004) Cavendish banana cultivars resistant to *Fusarium* wilt acquired through somaclonal variation in Taiwan. Plant Dis 88:580–588

81. Iacona C, Cirilli M, Zega A, Frioni A, Silvestri C, Muleo R (2013) A somaclonal myrobalan rootstock increases waterlogging tolerance to peach cultivar in controlled conditions. Sci Hortic 156:1–8

82. Israeli Y, Lahav E, Reuveni O (1995) *In vitro* culture of bananas. In: Gowen S (ed) Bananas and plantains. Chapman and Hall, London, pp 147–178

83. Iuliana C, Cerasela P (2014) The effect of the ultraviolet radiation on the somaclonal variability for *Solanum tuberosum*. Rom Biotechnol Lett 19(3):9339–9344

84. Jain SM (1997) Micropropagation of selected somaclones of *Begonia* and *Saintpaulia*. J Biosci 22:582–592

85. Jain SM (2001) Tissue culture-derived variation in crop improvement. Euphytica 118:153–166

86. Jaligot E, Beule T, Baurens FC, Billotte N, Rival A (2004) Search for methylation-sensitive amplification polymorphisms associated with the 'mantled' variant phenotype in oil palm (*Elaeis guineensis* Jacq.). Genome 47:224–228

87. Jaligot E, Adler S, Debladis É, Beulé T, Richaud F, Ilbert P, Finnegan EJ, Rival A (2011) Epigenetic imbalance and the floral developmental abnormality of the in vitro-regenerated oil palm *Elaeis guineensis*. Ann Bot 108:1453–1462

88. Jaya RS, Rao PS, Mhatre M (2002) Suitability of RAPD for analyzing spined and spineless variant regenerants of pineapple (*Ananas comosus* L., Merr.). Plant Mol Biol Rep 20:307

89. Jevremović S, Subotić A, Miljković D, Trifunović M, Petrić M, Cingel A (2012) Clonal fidelity of *Chrysanthemum* cultivars after long term micropropagation by stem segment culture. Acta Hortic 961:211–216

90. Joshi P, Dhawan V (2007) Assessment of genetic fidelity of micropropagated *Swertia chirayita* plantlets by ISSR marker assay. Biol Plant 51:22–26

91. Joyce SM, Cassells AC, Jain SM (2003) Stress and aberrant phenotypes in vitro culture. Plant Cell Tissue Organ Cult 74:103–121

92. Kaeppler SM, Kaeppler HF, Rhee Y (2000) Epigenetic aspects of somaclonal variation in plants. Plant Mol Biol 43:179–188

93. Kaity A, Ashmore SE, Drew RA (2009) Field performance evaluation and genetic integrity assessment of cryopreserved papaya clones. Plant Cell Rep 28:1421–1430

94. Kar B, Kuanar A, Singh S, Mohanty S, Joshi RK, Subudhi E, Nayak S (2014) In vitro induction, screening and detection of high essential

oil yielding somaclones in turmeric (*Curcuma longa* L.). Plant Growth Regul 72(1):59–66

95. Karp A (1992) The role of growth regulators in somaclonal variation. Br Soc Plant Growth Regul Annu Bull 2:1–9

96. Karp A (1994) Origins, causes and uses of variation in plant tissue cultures. In: Vasil IK, Thorpe TA (eds) Plant cell and tissue culture. Kluwer Academic Publishers, Dordrecht, pp 139–152

97. Karp A (1995) Somaclonal variation as a tool for crop improvement. Euphytica 85:295–302

98. Karp A (2000) Molecular tools for detecting genetic diversity. Acta Hortic 530:17–29

99. Khan S, Saeed B, Kauser N (2011) Establishment of genetic fidelity of in vitro raised banana plantlets. Pak J Bot 43:233–242

100. Khan SA, Rahman LU, Shanker K, Singh M (2014) *Agrobacterium tumefaciens*-mediated transgenic plant and somaclone production through direct and indirect regeneration from leaves in *Stevia rebaudiana* with their glycoside profile. Protoplasma 251(3):661–670

101. Kothari SL, Joshi A, Kachhwaha S, Ochoa-Alejo N (2010) Chilli peppers—a review on tissue culture and transgenesis. Biotechnol Adv 28(1):35–48

102. Krikorian AD, Irizarry H, Cronauer-Mitra SS, Rivera E (1993) Clonal fidelity and variation in plantain (*Musa* AAB) regenerated from vegetative stem and floral axis tips in vitro. Ann Bot 71:519–535

103. Krishna H, Singh SK (2007) Biotechnological advances in mango (*Mangifera indica* L.) and their future implication in crop improvement—a review. Biotechnol Adv 25:223–243

104. Krishna H, Singh D (2013) Micropropagation of *lasora* (*Cordia myxa* Roxb.). Indian J Hortic 70:323–327

105. Krishna H, Singh SK, Sharma RR, Khawale RN, Minakshi Patel VB (2005) Biochemical changes in micropropagated grape (*Vitis vinifera* L.) plantlets due to arbuscular-mycorrhizal fungi (AMF) inoculation during *ex vitro* acclimatization. Sci Hortic 106:554–567

106. Krishna H, Sairam RK, Singh SK, Patel VB, Sharma RR, Grover M, Nain L, Sachdeva A (2008) Mango explants browning: effect of ontogenic age: mycorrhization and pre-treatments. Sci Hortic 118:132–138

107. Kuanar A, Nayak PK, Subudhi E, Nayak S (2014) In vitro selection of turmeric somaclone resistant to *Fusarium oxysporum* f. sp. Zingiberi. Proc Natl Acad Sci India Sect B Biol Sci 84:1077–1082

108. Kukreja AK, Dhawan OP, Mathur AK, Ahuja PS, Mandal S (1991) Screening and evaluation of agronomically useful somaclonal variations in Japanese mint (*Mentha arvensis* L.). Euphytica 53:183–191

109. Kukreja AK, Dhawan OP, Ahuja PS, Sharma S, Kumar S (2000) Yield potential and stability behaviour of in vitro derived somaclones of Japanese mint (*Mentha arvensis* L.) under different environments. J Genet Breed 54(2):109–115

110. Kuksova VB, Piven NM, Gleba YY (1997) Plant cell variation and in vitro induced mutagenesis in grapevine. Plant Cell Tissue Organ Cult 49:17–27

111. Kuznetsova OI, Ash OA, Gostimsky SA (2006) The effect of the duration of callus culture on the accumulation of genetic alternation in pea *Pisum sativum* L. Russ J Genet 42:555–562

112. Larkin P, Scowcroft W (1981) Somaclonal variation—a novel source of variability from cell cultures for plant improvement. Theor Appl Genet 60:197–214

113. Lee SY, Su YU, Chou CS, Liu CC, Chen CC, Chao CP (2011) Selection of a new somaclone cultivar ‘Tai-Chiao No. 5’ (AAA,

Cavendish) with resistance to *Fusarium* wilt of banana in Chinese Taipei. Acta Hortic 897:391–397

114. Lestari EG (2006) In vitro selection and somaclonal variation for biotic and abiotic stress tolerance. Biodiversitas 7:297–301

115. Leva AR, Petruccelli R, Rinaldi LMR (2012) Somaclonal variation in tissue culture: a case study with olive. In: Leva AR, Rinaldi LMR (eds) Recent advances in plant in vitro culture. INTECH Open Access Publisher, Croatia, pp 123–150

116. Li R, Qu R, Bruneau AH, Livingston DP (2010) Selection for freezing tolerance in St. Augustine grass through somaclonal variation and germplasm evaluation. Plant Breed 129:417–421

117. Linacero R, Rueda J, Esquivel E, Bellido A, Domingo A, Vázquez AM (2011) Genetic and epigenetic relationship in rye, *Secale cereale* L., somaclonal variation within somatic embryo-derived plants. In Vitro Cell Dev Biol Plant 47:618–628

118. Litz RE, Mathews WH, Hendrix RC, Yurgalevitch C (1991) Mango somatic cell genetics. Acta Hortic 291:133–140

119. LoSchiavo F, Pitto L, Giuliano G, Torti G, Nuti-Ronchi V, Marazziti D, Vergara R, Orselli S, Terzi M (1989) DNA methylation of embryogenic carrot cell cultures and its variations as cased by mutation, differentiation, hormones and hypomethylating drugs. Theor Appl Genet 77:325–331

120. Lucia G, Castiglione MR, Turrini A, Ronchi VN, Geri C (2011) Cytogenetic and histological approach for early detection of "mantled" somaclonal variants of oil palm regenerated by somatic embryogenesis: first results on the characterization of regeneration system. Caryologia 64(2):223–234

121. Mallaya NP, Ravishankar GA (2013) In vitro propagation and genetic fidelity study of plant regenerated from inverted hypocotyl explants of eggplant (*Solanum melongena* L.) cv. Arka Shirish. 3 Biotech 3(1):45–52

122. Marino G, Beghelli S, Rombolà AD, Cabrini L (2000) *In vitro* performance at high culture pH and *in vitro* responses to Fe-deficiency of leaf-derived quince BA29 (*Cydonia oblonga*) somaclones regenerated at variable pH. J Hortic Sci Biotechnol 75:433–440

123. Martin K, Pachathundikandi S, Zhang C, Slater A, Madassery J (2006) RAPD analysis of a variant of banana (*Musa* sp.) cv. grande naine and its propagation via shoot tip culture. In Vitro Cell Dev Biol Plant 42:188–192

124. Martins M, Sarmento D, Oliveira MM (2004) Genetic stability of micropropagated almond plantlets, as assessed by RAPD and ISSR markers. Plant Cell Rep 23(7):492–496

125. Mathur AK (2010) Biotechnological studies in *Cymbopogons*. In: Akhila A (ed) Essential oil-bearing grasses: the genus *Cymbopogon*. CRC Press, Taylor & Francis Group, LLC, Boca Raton, pp 107–133

126. Mathur S (2013) Conservation of biodiversity through tissue culture. Res Rev J Microbiol Biotechnol 2:1–6

127. Mathur AK, Ahuja PS, Pandey B, Kukreja AK, Mathur A (1988) Development of superior strains of an aromatic grass—*Cymbopogon winterianus* Jowitt—through somaclonal variation. In: International conference on research in plant science and its relevance to future, Abstract, University Press, Delhi, p 160

128. Matsuda S, Sato M, Ohno S (2014) Cutting leaves and plant growth regulator application enhance somaclonal variation induced by transposition of VGs1 of *Saintpaulia*. J Jpn Soc Hortic Sci 83(4):308–316CrossRef

129. Menéndez-Yuffá A, Barry-Etienne D, Bertrand B, Georget F, Etienne H (2010) A comparative analysis of the development and quality of nursery plants derived from somatic embryogenesis and from seedlings for large-scale propagation of coffee (*Coffea arabica* L.). Plant Cell Tissue Organ Cult 102:297–307

130. Metzker ML (2010) Sequencing technologies—the next generation. Nat Rev 2010(1):31–46

131. Miguel C, Marum L (2011) An epigenetic view of plant cells cultured in vitro somaclonal variation and beyond. J Exp Bot 62:3713–3725

132. Miler N, Zalewska M (2014) Somaclonal variation of chrysanthemum propagated in vitro from different explants types. Acta Sci Pol Hortorum Cultus 13(2):69–82

133. Minerva G, Kumar S (2013) Micropropagation of gerbera (*Gerbera jamesonii* Bolus). In: Lambardi M, Ozudogru EA, Jain SM (eds) Protocols for micropropagation of selected economically-important horticultural plants. Methods Mol Biol 994:305–316

134. Miyao A, Nakagome M, Ohnuma T, Yamagata H, Kanamori H, Katayose Y, Takahashi A, Matsumoto T, Hirochika H (2012) Molecular spectrum of somaclonal variation in regenerated rice revealed by whole-genome sequencing. Plant Cell Physiol 53:256–264

135. Modgil M, Guleria N, Ghani M, Sharma JN (2012) Identifying somaclonal variants of the apple rootstock Malling 7 resistant to white root rot. Sci Hortic 137:148–155

136. Mohanty S, Parida R, Singh S, Joshi RK, Subudhi E, Nayak S (2011) Biochemical and molecular profiling of micropropagated and conventionally grown *Kaempferia galanga*. Plant Cell Tissue Organ Cult 106:39–46

137. Mondal TK, Bantawa P, Sarkar B, Ghosh P, Chand PK (2013) Cellular differentiation, regeneration, and secondary metabolite production in medicinal *Picrorhiza* spp. Plant Cell Tissue Organ Cult 112(2):143–158

138. Morozova O, Marra MA (2008) Applications of next-generation sequencing technologies in functional genomics. Genomics 92:255–264

139. Morrison RA, Loh WHT, Green SK, Griggs TD, McLean BT (1989) Tissue culture of tomato and pepper: new tools for plant breeding. Tomato and pepper production in the tropics. In: Proceedings of international symposium on integrated management practices, Tainen, Taiwan. Shanhua, Taiwan, AVRDC, pp 44–50

140. Mujib A, Banerjee S, Dev Ghosh P (2007) Callus induction, somatic embryogenesis and chromosomal instability in tissue culture-raised hippeastrum (*Hippeastrum hybridum* cv. United Nations). Propag Ornam Plants 7:169–174

141. Muleo R, Iacona C, Pistelli L, Loreti F (2006) A novel Mr. S. 2/5 peach rootstock clone tolerant to flooding stress. Adv Hortic Sci 20:208–214

142. Munir F, Naqvi SMS, Mahmood T (2011) *In vitro* culturing and assessment of somaclonal variation of *Solanum tuberosum* var. desiree. Turk J Biochem 36:296–302

143. Murti RH, Kim HY, Yeoung YR (2013) Effectiveness of gamma ray irradiation and ethyl methane sulphonate on in vitro mutagenesis of strawberry. Afr J Biotechnol 12:4803–4812

144. Nacheva LR, Gercheva PS, Andonova MY, Panayotova DV, Dzhuvinov VT (2014) Somaclonal variation: a useful tool to improve disease resistance of pear rootstock ‘Old Home × Farmingdale’ (OHF 333) (*Pyrus communis* L.). Acta Hortic 1056:253–258

145. Naseer S, Mahmood T (2014) Tissue culture and genetic analysis of somaclonal variations of *Solanum melongena* L. cv. Nirrala. Cent Eur J Biol 9(12):1182–1195

146. Nassar AM, Abdulnour J, Leclerc YN, Li XQ, Donnelly DJ (2011) Intraclonal selection for improved processing of NB ‘Russet Burbank’potato. Am J Potato Res 88(5):387–397

147. Nassar AM, Kubow S, Leclerc YN, Donnelly DJ (2014) Somatic mining for phytonutrient improvement of ‘Russet Burbank’ potato. Am J Potato Res 91:89–100

148. Nayak S, Debata BK, Srivastava VK, Sangwan NS (2003) Evaluation of agronomically useful somaclonal variants in Jamrosa (a hybrid *Cymbopogon*) and detection of genetic changes through RAPD. Plant Sci 164:1029–1035

149. Nayak S, Kaur T, Mohanty S, Ghosh G, Choudhury R, Acharya L, Subudhi E (2010) *In vitro* and *ex vitro* evaluation of long-term micropropagated turmeric as analyzed through cytophotometry, phytoconstituents, biochemical and molecular markers. Plant Growth Regul 64:91–98

150. Ngezahayo F, Dong Y, Liu B (2007) Somaclonal variation at the nucleotide sequence level in rice (*Oryza sativa* L.) as revealed by RAPD and ISSR markers, and by pairwise sequence analysis. J Appl Genet 48:329–336

151. Nhut DT, Hai NT, Thu PTM, Thi NN, Hien TTD, Tuan TT, Nam NB, Huy NP, Chien HX, Jain SM (2013) Protocol for inducing flower color somaclonal variation in torenia (*Torenia fournieri* Lind.). Methods Mol Biol 11013:455–462

152. Nivas SK, DSouza L (2014) Genetic fidelity in micropropagated plantlets of *Anacardium occidentale* L. (Cashew) an important fruit tree. Int J Sci Res 3:2142–2146

153. Nwauzoma AB, Jaja ET (2013) A review of somaclonal variation in plantain (*Musa* spp): mechanisms and applications. J Appl Biosci 67:5252–5260

154. Ooi SE, Novák O, Doležal K, Ishak Z, Ong-Abdullah M (2013) Cytokinin differences in in vitro cultures and inflorescences from normal and mantled oil palm (*Elaeis guineensis*Jacq.). J Plant Growth Regul 32(4):865–874

155. Opabode JT, Adebooye OC (2005) Application of biotechnology for the improvement of Nigerian indigenous leaf vegetables. Afr J Biotechnol 4(3):138–142

156. Panda MK, Mohanty S, Subudhi E, Acharya L, Nayak S (2007) Assessment of genetic stability of micropropagated plants of

*Curcuma longa* L. by cytophotometry and RAPD analysis. Int J Integr Biol 1(3):189–195

157. Parida R, Mohanty S, Nayak S (2013) In vitro propagation of *Hedychium coronarium*Koen. through axillary bud proliferation. Plant Biosyst 147(4):905–912

158. Pathak H, Dhawan V (2012) ISSR assay for ascertaining genetic fidelity of micropropagated plants of apple rootstock Merton 793. In Vitro Cell Dev Biol Plant 48:137–143

159. Patnaik J, Sahoo S, Debata BK (1999) Somaclonal variation in cell suspension culture-derived regenerants of *Cymbopogon martinii* (Roxb.) Wats var. Motia. Plant Breed 118:351–354

160. Patzak J (2003) Assessment of somaclonal variability in hop (*Humulus lupulus* L.) in vitro meristem cultures and clones by molecular methods. Euphytica 131(3):343–350

161. Pérez G, Yanez E, Isidrón M, Lorenzo JC (2009) Phenotypic and AFLP characterization of two new pineapple somaclones derived from in vitro culture. Plant Cell Tissue Organ Cult 96:113–116

162. Pérez G, Mbogholi A, Sagarra F, Aragón C, González J, Isidrón M, Lorenzo JC (2011) Morphological and physiological characterization of two new pineapple somaclones derived from in vitro culture. In Vitro Cell Dev Biol Plant 47:428–433

163. Pérez G, Yanez E, Mbogholi A, Valle B, Sagarra F, Yabor L, Aragón C, González J, Isidrón M, Lorenzo JC (2012) New pineapple somaclonal variants: P3R5 and dwarf. Am J Plant Sci 3:1–11

164. Phillips RL, Kappler SM, Olhof R (1994) Genetic instability of plant tissue cultures: breakdown of normal controls. Proc Natl Acad Sci USA 91:5222–5226

165. Piagnani MC, Chiozzotto R (2010) Shoot regeneration, in vitro performances of regenerated shoots and transient expression in morphogenic explants in *Prunus avium*cultivar ‘Burlat C1’. Eur J Hortic Sci 3:132–138

166. Pijut PM, Beasley RR, Lawson SS, Palla KJ, Stevens ME, Wang Y (2012) In vitro propagation of tropical hardwood tree species—a review (2001–2011). Propag Ornam Plants 12(1):25–51

167. Podwyszynska M, Kuras A, Korbin M (2010) Somaclonal variation in micropropagated tulips as a source of novel genotypes—field and molecular characteristic. Acta Hortic 855:225–232

168. Prado MJ, Gonzalez MV, Romo S, Herrera MT (2007) Adventitious plant regeneration on leaf explants from adult male kiwifruit and AFLP analysis of genetic variation. Plant Cell Tissue Organ Cult 88(1):1–10

169. Purwati RD, Sudarsono S (2007) Resistance of Abaca somaclonal variant against*Fusarium*. Hayati J Biosci 14:133–139

170. Rabiei K, Khodambashi N, Omidi M (2011) Use of somaclonal variation on improvement of drought resistant lines of carrot (*Daucus carota* L.). J Hortic Sci (Agric Sci Technol) 25:156–169

171. Rana S, Dhar N, Bhat WW, Razdan S, Khan S, Dhar RS, Dutt P, Lattoo SK (2012) A 12-deoxywithastramonolide-rich somaclonal variant in *Withania somnifera* (L.) Dunal—molecular cytogenetic analysis and significance as a chemotypic resource. In Vitro Cell Dev Biol Plant 48(5):546–554

172. Ravindra NS, Ramesh SI, Gupta MK, Jhang T, Shukla AK, Darokar MP, Kulkarni RN (2012) Evaluation of somaclonal variation for genetic improvement of patchouli (*Pogostemon patchouli*), an exclusively vegetatively propagated aromatic plant. J Crop Sci Biotechnol 15:33–39

173. Ray T, Dutta I, Saha P, Das S, Roy SC (2006) Genetic stability of three economically important micropropagated banana (*Musa* spp.) cultivars of lower Indo-Gangetic plains, as assessed by RAPD and ISSR markers. Plant Cell Tissue Organ Cult 85(1):11–21

174. Resende GA, Mangolin CA, Machado MDF (2010) Somaclonal variation in *Cereus peruvianus* Mill. (Cactaceae): its potential to

generate new varieties and broaden the species's genetic basis. J Basic Appl Genet 21:33–42

175. Rival A, Ilbert P, Labeyrie A, Torres E, Doulbeau S, Personne A, Dussert S, Beulé T, Durand-Gasselin T, Tregear JW, Jaligot E (2013) Variations in genomic DNA methylation during the long-term in vitro proliferation of oil palm embryogenic suspension cultures. Plant Cell Rep 32:359–368

176. Rosati P, Mezzetti B, Ancherani M, Foscolo S, Predieri S, Fasolo F (1990) *In vitro* selection of apple rootstock somaclones with *Phytophthora cactorum* culture filtrate. Acta Hortic 280:409–416

177. Sahijram L, Soneji J, Bollamma K (2003) Analyzing somaclonal variation in micropropagated bananas (*Musa* spp.). In Vitro Cell Dev Biol Plant 39:551–556

178. Sakhanokho HF, Witcher AL, Pounders CT, Spiers JM (2012) 'Ramata': a new dwarf variegated *Hedychium* (ornamental ginger) cultivar. HortSci 47:803–805

179. Sales EK, Butardo NG (2014) Molecular analysis of somaclonal variation in tissue culture derived bananas using MSAP and SSR markers. Int J Biol Vet Agric Food Eng 8:63–610

180. Sanputawong S, Te-Chato S (2011) Analysis of somaclonal variation of callus, somatic embryo and plant regeneration of in vitro oil palm (*Elaeis guineensis* Jacq.). J Agric Technol 7(2):531–545

181. Saravanan S, Sarvesan R, Vinod MS (2011) Identification of DNA elements involved in somaclonal variants of *Rauvolfia serpentina* (L.) arising from indirect organogenesis as evaluated by ISSR analysis. Indian J Sci Technol 4:1241–1245

182. Sato M, Kawabe T, Hosokawa M, Tatsuzawam F, Doi M (2011a) Tissue culture induced flower-color changes in Saintpaulia caused by excision of the transposon inserted in the flavonoid 39, 59 hydroxylase (F3959H) promoter. Plant Cell Rep 30:929–939

183. Sato M, Hosokawa M, Doi M (2011b) Somaclonal variation is induced de novo via the tissue culture process: a study quantifying mutated cells in Saintpaulia. PLoS ONE 6:e23541.

184. Saxena G, Verma PC, Rahman LU, Banerjee S, Kumar S (2008) Field performance of somaclones of rose scented geranium (*Pelargonium graveolens* L' Her Ex Ait.) for evolution of their essential oil yield and composition. Ind Crops Prod 27:86–90

185. Schellenbaum P, Mohler V, Wenzel G, Walter B (2008) Variation in DNA methylation patterns of grapevine somaclones (*Vitis vinifera* L.). BMC Plant Biol 8:78

186. Sharma S, Bryan G, Winfield M, Millam S (2007) Stability of potato (*Solanum tuberosum*L.) plants regenerated via somatic embryos, axillary bud proliferated shoots, microtubers and true potato seeds: a comparative phenotypic, cytogenetic and molecular assessment. Planta 226:1449–1458

187. Shearman JR, Jantasuriyarat C, Sangsrakru D, Yoocha T, Vannavichit A, Tragoonrung S, Tangphatsornruang S (2013) Transcriptome analysis of normal and mantled developing oil palm flower and fruit. Genomics 101(5):306–312

188. Shen X, Chen J, Kane M, Henny R (2007) Assessment of somaclonal variation in*Dieffenbachia* plants regenerated through indirect shoot organogenesis. Plant Cell Tissue Organ Cult 91:21–27

189. Siragusa M, Carra A, Salvia L, Puglia A, De Pasquale F, Carimi F (2007) Genetic instability in calamondin (*Citrus madurensis* Lour.) plants derived from somatic embryogenesis induced by diphenylurea derivatives. Plant Cell Rep 26:1289–1296

190. Skirvin RM, Janick J (1976) Tissue culture-induced variation in scented *Pelargonium* spp. J Am Soc Hortic Sci 101:281–290

191. Slazak B, Sliwinska E, Saługa M, Ronikier M, Bujak J, Słomka A, Göransson U, Kuta E (2015) Micropropagation of *Viola uliginosa* (Violaceae) for endangered species conservation and for somaclonal

variation-enhanced cyclotide biosynthesis. Plant Cell Tissue Organ Cult 120:179–190

192. Smith MK (1988) A review of factors influencing the genetic stability of micropropagated bananas. Fruits 43:219–223

193. Smulders M, de Klerk G (2011) Epigenetics in plant tissue culture. Plant Growth Regul 63:137–146

194. Stanišić M, Raspor M, Ninković S, Milošević S, Ćalić D, Bohanec B, Trifunovića M, Petrića M, Subotić A, Jevremović S (2015) Clonal fidelity of *Iris sibirica* plants regenerated by somatic embryogenesis and organogenesis in leaf-base culture—RAPD and flow cytometer analyses. S Afr J Bot 96:42–52

195. Sun S, Zhong J, Li S, Wang X (2013) Tissue culture-induced somaclonal variation of decreased pollen viability in torenia (*Torenia fournieri* Lind.). Bot Stud 54(1):36

196. Takahashi H, Matsumoto T, Takai T (1993) Somaclonal variants from strawberry cv. Morioka-16, M16-AR 1, 2 and 3 resistant to several isolates of *Alternaria alternata*strawberry pathotype occurring in Tohoku and Hokkaido [Japan]. J Jpn Soc Hortic Sci 61:821–826

197. Tang CY, Liu CC, Hwang SC (2000) Improvement of the horticultural traits of Cavendish banana (Musa spp., AAA group I). Selection and evaluation of a semi-dwarf clone resistant to *Fusarium* wilt. J Chin Soc Hortic Sci 46:173–182

198. Tanurdzic M, Vaughn MW, Jiang H, Lee TJ, Slotkin RK, Sosinski B, Thompson WF, Doerge RW, Martienssen RA (2008) Epigenomic consequences of immortalized plant cell suspension culture. PLoS Biol 6(12):2880–2895

199. Thieme R, Griess H (2005) Somaclonal variation in tuber traits of potato. Potato Res 48:153–165

200. Tican A, Câmpeanu G, Chiru N, Ivanovici D (2008) Using of unconventional methods for obtaining somaclonal variations, having

as goal making of new potato varieties with resistance at diseases and pests. Rom Biotechnol Lett 13:3791–3798

201. Tiwari JK, Chandel P, Gupta S, Gopal J, Singh BP, Bhardwaj V (2013) Analysis of genetic stability of in vitro propagated potato microtubers using DNA markers. Physiol Mol Biol Plants 19:587–595

202. Toyoda H, Horikoshi K, Yamano Y, Ouchi S (1991) Selection of *Fusarium* wilt disease resistance from regenerant derived from callus of strawberry. Plant Cell Rep 10:167–170

203. Vázquez AM (2001) Insight into somaclonal variation. Plant Biosyst 135:57–62

204. Veitia-Rodriguez N, Francisco-Cardoso J, Perez JN, Garcia-Rodriguez L, Bermudez-Caraballosos I, Garcia-Rodriguez L, Padron-Montesinos Y, Orellana-Perez P, Romero-Quintana C, Hernandez N (2002) Evaluations in field of somaclones of Irish potatoes (*Solanum tuberosum* Lin.) of the variety Desiree obtained by somaclonal variation and in vitro mutagenesis. Biotechnol Veg 2:21–26

205. Vidal BC, Mello MLS, Illg RD, de Campos Vidal B, Campos De, Vidal B (1993) Chromosome number and DNA content in cells of a biotechnologically selected somaclone of garlic (*Allium sativum* L). Rev Bras Genet 16(2):347–356

206. Viseur J (1990) Evaluation of fire blight resistance of somaclonal variants obtained from the pear cultivar 'Durondeau'. Acta Hortic 273:275–284

207. Vujović T, Ružić Đ, Cerović R, Momirović GŠ (2010) Adventitious regeneration in blackberry (*Rubus fruticosus* L.) and assessment of genetic stability in regenerants. Plant Growth Regul 61:265–275

208. Wacksman JT (1997) DNA methylation and the association between genetic and epigenetic changes: relation to carcinogenesis. Mutat Res 375:1–8

209. Whitehouse AB, Johnson AW, Passey AJ, McLeary KJ, Simpson DW (2014) Serenity': a paler skin-coloured somaclonal variant of the short-day cultivar 'Florence. Acta Hortic 1049:819–821

210. Xu M, Li X, Korban SS (2004) DNA-methylation alterations and exchanges during in vitro cellular differentiation in rose (*Rosa hybrida* L.). Theor Appl Genet 109:899–910

211. Yang H, Schmidt H (1994) Selection of a mutant from adventitious shoots formed in X ray treated cherry leaves and differentiation of standard and mutant with RAPDs. Euphytica 77:89–92

212. Yusnita Y, Widodo W, Sudarsono S (2005) In vitro selection of peanut somatic embryos on medium containing culture filtrate of *Sclerotium rolfsii* and plantlet regeneration. HAYATI J Biosci 12(2):50–56

213. Zayova E, Vassilevska IR, Kraptchev B, Stoeva D (2010) Somaclonal variations through indirect organogenesis in eggplant (*Solanum melongena* L.). Biol Divers Conserv 3:1–5

214. Zebrowska JI (2010) In vitro selection in resistance breeding of strawberry (*Fragaria* x*ananassa* Duch.). Commun Agric Appl Biol Sci 75:699–704

215. Zhang LQ, Cheng ZH, Khan MA, Zhou YL (2012) In vitro selection of resistant mutant garlic lines by using crude pathogen culture filtrate of *Sclerotium cepivorum* Australas. Plant Pathol 41:211–217

# CHAPTER 10

# Land use and land cover (LULC) changes in semi-arid sub-watersheds of Laikipia and Athi River basins, Kenya, as influenced by expanding intensive commercial horticulture

**Faith K. Muriithi**[1]

Department of Earth and Environmental Studies,
Montclair State University, Montclair, NJ 07043, USA

## HIGHLIGHTS

- Land transformation occurred at varying magnitudes during the 25 years period.
- Woody grasslands cover vast areas of in Laikipia and Athi River.
- Laikipia, had highest percentage annual rate of change to urban (20.2%).
- Athi River had highest percentage annual rate of change to agriculture (37.7%).
- Observed land transformations largely attributed to socio-economic drivers.

# ABSTRACT

Agriculturally productive watersheds in Kenya are experiencing important land transformations with consequences on environmental resources. Previous work identified spatial pockets of vegetation decline across sub-watersheds with intensive farming. While this is crucial information in delineating impacted regions, it is important to understand the type of occurring changes, spatial patterns and the rate of change. The study focused in the Athi River and Laikipia regions which have intensified commercial horticulture, yet marginalized semi-arid to arid environments. The study quantified land use and land cover changes between 1984 and 2009/2010, identifying areas of change and the average annual rate of change. Landsat (5 TM, MSS and 7 ETM+) multitemporal data were used in classification and change detection analysis. From the results, both regions experienced intense land transformation at varying magnitudes during the 25 years period. Woody grasslands covered vast areas. In Laikipia, the averaged percentage annual rate of land change to settlements/urban use was highest (20.2%). There was a decline in barren area, forest and wooded grasslands (−1.4%, −0.7% and −0.3% respectively). In the Athi River region, the averaged percentage annual rate of changes was in the order of agriculture, water, settlement, bare soils, woody grasslands (37.7%, 17.2%, 8.4%, 3.9, 2.9% respectively) while dark bare soils, and forest cover declined (−3.4% and −3.3% respectively). Observed LULC transformations were largely attributed to socio-economic drivers including increased human migration into agriculturally productive sub-watersheds fueling growth of settlement, increased irrigation farming and therefore increased area under water, and growth of cities and industrial areas.



# KEYWORDS

Land transformation; Deforestation; Horticulture; Population

# 1. INTRODUCTION

Land use and land cover (LULC) transformation is an important research item due to the increasing impacts on ecosystems and link to global, regional and local climate change and variability (Bajocco et al., 2012, Liu et al., 2014, Mahmood et al., 2006, Turner et al., 2007 and West et al., 2010). The common drivers of LULC include agricultural encroachment, deforestation, road construction, dams and irrigation, wetland modification, mining, expansion of urban environments (Liu et al., 2014), and coastal zone degradation (Patz and Olson, 2008). At the watershed scale, land cover changes pose risks of water pollution due to loss of vegetative cover, which acts as a barrier to the movement of materials into water systems, reducing runoff (Notter et al., 2007 and Schneider and Gil Pontius, 2001). This significantly increases possible impacts on terrestrial and aquatic systems, and social-economics of communities (Adger et al., 2005).

In Kenya, the proliferation of intensive commercial horticulture in Kenya in the recent past has influenced LULC changes in central districts that are agriculturally productive (Becht et al., 2005 and Francis, 2014). The uneven clustering of large scale farms, over-extraction of surface water for irrigation, clearing of marginal lands and continuous tilling to increase year round production of fresh produce are some activities impacting the use and management of environmental resources (Aeschbacher et al., 2005 and Owiti and Oswe, 2007). The horticulture sector is labor intensive, attracting population into sub-watersheds, leading to high population density and growth of unplanned settlement (Muriuki et al., 2011). Unrestrained growth within sub-watersheds exerts pressure on environmental resources, enhances degradation and competition between and within species (Barrow, 2006), factors that can augment watersheds' vulnerability to climate change effects.

Population increase exceeding the carrying capacity of a region is destructive, often causing an over burden on other species within those areas which can result to reduced biodiversity (Oehl et al., 2003). Studies show that population and human settlement have a strong negative correlation with net primary productivity (NPP), with trends of NPP shown to decrease as settlement, or population increases (Haberl et al., 2007 and Lu et al., 2010). Recent regional observation show that areas experiencing reduced main growing season precipitation in East Africa coincidentally experienced increasing population pressures, and vegetation browning (Pricope et al., 2013). Even so, the drying precipitation patterns only partially explained the observed vegetation browning trends. The study pointed to other factors such as population increase and land-use changes as responsible for the decline in vegetation condition.

The drivers of land transformation are complex, and region dependent, requiring area based studies to investigate amount and rate of changes (Kasperson et al., 1996). Attempts to understand LULC changes in Kenya provide mixed conclusions on the nature, extent and impacts on ecosystems. Baldyga et al. (2008), show that the Mau forest underwent major changes after 1995, due to deforestation. Vast forest cover was converted to mixed small-scale agriculture and managed pastures, resulting in spatially variable landscapes theorized to have significant impacts on ecological and hydrologic systems in the area. In Laikipia district (a semi arid to arid region) rapid changes in land ownership and land use are linked to population increase (Mutiga et al., 2010) as a result of augmented regional intensive horticulture (Mungai et al., 2004). Similar observations are noted in the Athi River region (Kithiia, 2012a), and also related to intensive agriculture, industrial activity and population pressure, all which influence surface water quality in the basin and therefore the need for remedial actions. These findings are important in understanding the impacts of increasing LULC changes on natural resources in the region. However, the dynamics of land use conversions and rate of change between uses, which provide more specific indication of for example the sources and potential pathways of pollutant/sediment transport within watersheds, remain un quantified. Areal numbers provide magnitude of change, providing information necessary in designing practical management strategies to ease pressure.

The research to monitor, quantify and understand environmental consequences of changes in earth systems at spatiotemporal scales has widely utilized remote sensing data (Giri et al., 2005 and Yang et al., 2013). More so, there is a growing interest in research that integrates remote sensing information with socio economic data to seek resource management solutions (Liverman et al., 1998). Satellite data provides faster and consistent means of studying terrestrial surface and changes over time, enabling robust, efficient and cost-effective method to map biomass distribution (Jensen, 2005, Lu et al., 2004 and Maingi and Marsh, 2001).

To this end, freely available Landsat data are commonly used in LULC change detection investigation. They provide a long history and higher frequency of archives, and continuous data that enable comprehensive assessments and mapping of LULC (Abd El-Kawy et al., 2011 and Wulder et al., 2008). In sub-Saharan Africa, several studies have utilized Landsat data. Kashaigili and Majaliwa (2010), mapped LULC changes in the Malagarasi river catchment in Tanzania. Their findings indicated significant land transformations linked to increased settlements and cultivation. Bakr et al. (2010), used multi-temporal Landsat data to monitor land cover changes in Egypt, noting that it provides an effective and precise estimation of human impact on the environment. In central Kenya, Imbernon (1999) assessed land-use changes in the upper Embu area, and found a minor loss of tree cover, but a high loss of bush land. Recently, Kirui et al. (2013) utilized Landsat data to map mangrove forest cover along the Kenyan coastline. They reported an annual loss of 18% (0.7% per year) in mangroves between 1985 and 2010.

This paper investigates the dynamics of land use and land cover transformations in two study regions characterized by increased intensive commercial horticulture and human pressure. These regions are within arid to semi-arid type of climate, which could increase their susceptibility to environmental degradation. The specific objectives were to develop maps of LULC change between 1984 and 2009/2010, to determine quantity of cover/use change, the annual rate of change, and the trends of population density in sub watersheds within the study zones. Data on acreage of land conversion over years and maps of change shed light into dimensions and probable drivers of change, providing a base for future studies. The study contributes information critical to designing improved watershed management plans and policies on resource allocation.

The succeeding sections of the study include; a detailed Section 2 which present a short description of the study regions, the available satellite data, and their processing; a section highlighting the results and discussions and finally a conclusion of the study and suggestions for improvement.

# 2. MATERIALS AND METHODS

## 2.1. Study areas

### 2.1.1. Laikipia study area

The study region comprised of 5 sub-watersheds, covering 6188 sq. km within the semi-arid to arid land of Ewaso Nyiro catchment (Fig. 1A). It mainly falls on the eastern escarpment of the Rift Valley, primarily occupied by extensive ranches, mixed small scale agriculture, vast savanna grasslands and woodland, forest cover surrounding Mt. Kenya, and large scale farms at the foot zone of Mount Kenya (Fig. 1A).

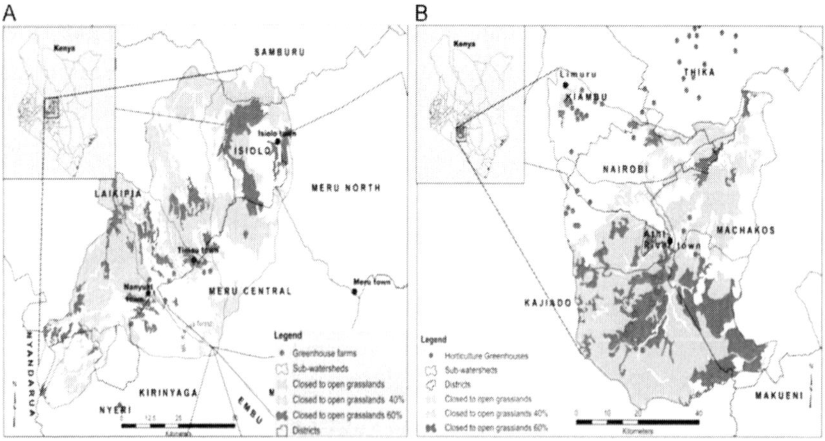


**Figure 1.** Maps showing the study areas with a grass cover basemap (*Source*: *World Resource Institute GIS data*). (A) Laikipia area encompasses 5 sub-watersheds; (B) Athi River area encompasses 4 sub-watersheds. Both regions have increased large-scale greenhouse horticulture, and vast grasslands compared to other cover types.

The 5 sub-watersheds had clusters of large-scale greenhouses for commercial horticulture (Justus and Yu, 2014). Settlements and urbanizing town centers are common with the remainder of land area occupied by wildlife habitat, forest reserves and conservation parcels. The rainfall is generally bimodal with an average annual rainfall of about 700 mm/year (Gichuki et al., 1998 and Thomas and Liniger, 1994).

The long rains occur from March to June, and short rains occur from October to December (Berger, 1989 and Notter, 2003). Between the two seasons, dry conditions persist influencing the type of vegetation cover to that of savanna ecosystems. Besides, the presence of Mt. Kenya at high elevation, form a highland-lowland system that prompts a striking climatic and ecological gradient, from humid moorlands and forests on the mountain slopes to arid Acacia bushland in the lowland, with a diverse pattern of land uses (Decurtins, 1992 and Mutiga et al., 2010). Cultivation occurs towards the mountain foot zones, giving way to savannah and acacia vegetation. The altitude in Laikipia gradually falls from about 2000 m to 1700 m, and becomes increasingly dry, with an average annual rainfall of between 450 mm and 700 mm. Most of Laikipia is located on the leeward slopes of Mt. Kenya, which makes the area exhibit arid and semi-arid climatic characteristics.

Soils are primarily deep vertisols (verto-luvic Phaozem) of dark gray clays, that are imperfectly to moderately drained, and developed from volcanic activity of Mt. Kenya (Liniger et al., 1998). These soils support little rain-fed agriculture, and limited land use due to their high erodibility (Weismann, 1998). The Ewaso Nyiro River is a main source of water towards the drier outer areas of the study zone, while Mt Kenya forms a water source for several rivers that support farming at the mountain foot zones.

### 2.1.2. Athi River region

This study region was a subset of 4 sub-watersheds covering 4199.2 sq. km within the greater Athi River catchment (Fig. 1B). Like Laikipia, shrubs and woody grasslands cover a vast area in this study region towards the southern arid and semi-arid zones of Kajiado district, where only 8% of the district land has potential to support rain-fed agriculture (De Leeuw et al., 1991). The soils are predominantly deep black Vertisols that support and maintain the rangelands and pastoralist activity. The upper catchment is shared by Nairobi County (at the core), and Kiambu county towards the northwest (Fig. 1B). In Kiambu, towns like Limuru and its surroundings are dominated by intensive agriculture that supports approximately 70% of the population (MoPND, 1994).

The rapid expansion of Nairobi city and its environs have a large influence on the natural landscape. Large tracks of forest land, mixed rangelands and bush lands (UNEP, 2008) have been converted to urban dwellings, roads and city infrastructure, causing receding forest cover. The demand for food by the proliferating population led to major changes in land use and cover patterns in the outskirts of the city, converting to agricultural more intensive commercial large scale horticulture in the marginalized areas of the Athi River basin (MDC, 1993 and Tibaijuka, 2007). The mean annual rainfall varies between 300 mm and 800 mm, and falls in the two main rainy seasons; October to December months (short rains) and March to May (long rains). This kind of annual rainfall is inadequate for rain-fed agriculture. The vegetation is predominantly open grasslands, that are distributed along altitude gradient of 1100–200 m (*Themeda* species), and at 450–1200 m (*Chloris* species) ( Rattray, 1960). Forest cover is minimal to the south, and where present, it is confined in isolated patches on hill tops and along the flat low lying phonolithic lavas ranges similar to those found in the Laikipia plateau.

The mix of prevailing land use types and the arid to semi-arid climatic conditions in the basin make it vulnerable to environmental resource degradation including high rates of soil erosion on the bare barren surfaces (Tiffen and Mortimore, 1992).

Both the Laikipia and Athi River basin study areas share almost similar environmental factors (aridity, low rainfall, vegetation gradients, rapid population expansion into marginal lands), and highly susceptible to climate changes and declining land productivity and therefore important case studies. Both regions have shown an increase in large-scale commercial greenhouses (Justus and Yu, 2014). Farming is expanding towards the more marginalized arid areas presenting a concern to the fragile ecosystems whose natural resources require better management.

# 3. METHODS

The study integrates both remote sensing and geographic information systems (GIS) techniques to investigate land use and land cover (LULC) dynamics in two study areas delineated as regions with increasing large scale intensive commercial horticulture production and rapid population growth. The study incorporates demographic information i.e. population density of the sub-watersheds derived from census data (1 and 1 and2009), and graphed to visualize trend in population, a prominent key driver of LULC changes.

## 3.1. Remote sensed data, processing and change detection

### 3.1.1. Image selection and acquisition

Landsat-5 Multispectral Scanner (MSS) and Thematic Mapper (TM) and Landsat-7 Enhanced Thematic Mapper (ETM+) images were obtained for the years 1984, 2009/2010 (Table 1), a period of ~25 years which provide ample time to monitor LULC change. An effort was made to acquire cloud free image scenes. It is easier to note phenology differences in vegetation during dry seasons than in wet vegetative seasons. In this study, the scenes utilized were acquired at dates that coincided with drier than normal conditions due to drought in the region in 1984, 2008, and in 2010 (Shisanya, 1990, Huho and Mugalavai, 2010 and Global Facility for Disaster Reduction and Recovery (GFDRR), 2012).

**Table 1.** Remotely sensed data used in the analysis of land use/cover change. TM=Thematic Mapper sensor; ETM+=Enhanced Thematic Mapper plus sensor.

| Study area | Satellite sensor | Path/Row | Acquisition date (m/d/yr) | Spatial resolution | Sun azimuth | Sun elevation | Season |
|---|---|---|---|---|---|---|---|
| Laikipia | Landsat5 MSS | P168/60 | 4/21/1984 | 60 | 69.63054 | 53.46713 | [a]Rain |
| | Landsat7 ETM+ | P168/60 | 1/12/2009 | 30 | 127.0982 | 52.38331 | Dry |
| Athi River | Landsat5 TM | P168/61 | 8/27/1984 | 30 | 71.22358 | 53.1759 | [b]Rain |
| | Landsat7 ETM+ | P168/61 | 12/17/2010 | 30 | 131.2164 | 55.23683 | Dry |

[a] It was a dry period due to the 1984 drought (Shisanya, 1990).
[b] It was a dry period due to the prolonged 2008–2010 drought (Huho and Mugalavi, 2010; GFDRR, 2012).

### 3.1.2. Image pre-processing and atmospheric correction of scenes

The Landsat images were downloaded from the U.S. Geological Survey (USGS) Earth Explorer website. The images were already geo-referenced to rectify for alignment, and co-registered to the Universal Transverse Mercator (UTM) projection system (zone: 37N, datum: WGS-84) avoiding the need for geo-rectification which would have employed ground control points (GCPs), paying attention to acquire the least root mean square error (RMSE) in pixels. The image layers were imported and layer stacked to form multispectral images. These steps were carried out in ERDAS Imagine 2014.

Remotely sensed satellite data are affected by artifacts such as haze resulting from water vapor and aerosol particles present in the atmosphere which influence the sensor signal. For land cover change detection using multidate satellite derived images, proper atmospheric correction is necessary because differences in the atmosphere for the two dates can present false indication of change or mask areas of real change (Richter, 1996 and Townshend, 1994). Impacts of atmospheric correction on image classification are evaluated (Kaufman, 1994). In the study, hazy conditions on scenes at the top of Mt. Kenya (5199 masl) and a prevailing rugged terrain can introduce varying scene illumination e.g. sunny, shady hills-conditions that modify the actual spectral behavior of cover as seen by the sensor. This necessitated atmospheric correction of the scenes to reduce significant errors in image classification. Besides, the use of multi-temporal images requires atmospheric correction of scenes before classification processes.

The study utilized ATmospheric and Topographic CORrection (ATCOR) terrain and flat modules with algorithms to eliminate haze, atmospheric and solar illumination effects. Atmospheric-topographic correction enables a comparison of multi-temporal scenes, as the per-pixel digital numbers recorded by the sensor can be transformed into surface reflectance values (Neubert and Meinel, 2005). The pre-processing steps equalize the effects of the sun position due to the recording time of day and the seasonal differences (Neubert and Meinel, 2005). Furthermore, through topographic/terrain correction, the noise resulting from the illumination effects caused by the relief patterns e.g. high relief areas and shadowing on hillsides can be minimized. For this reason, a high resolution digital elevation model (DEM) is essential.

To enable multi sensor scene comparison, the MSS images were re-sampled using a cubic convolution from the 60 m spatial resolution to 30 m. In the study, the ATCOR2 algorithm available in ERDAS Imagine was used for haze removal, atmospheric and topographic correction of the scenes. The algorithm utilized a Shuttle Radar Topographic Mission (SRTM) 90 m digital elevation model (DEM), to derive terrain files including slope, aspect, skyview and shadow files. The files are used in topographic normalization, which tries to account for the brightness distortion caused by differences in illumination due to the changing sun angle relative to the terrain angle i.e. slope, aspect (Smith et al., 1980). It is also shown to improve classification accuracy (Jensen, 1996). The process involved setting the primary scene parameters e.g. the acquisition date, the sensors name, the scene visibility and the solar zenith and azimuth. The scene information alongside the generated files were utilized during the topographic correction process (see Neubert and Meinel, 2005), resulting to corrected reflectance images that were used for classification.

### 3.1.3. Image classification

A supervised land cover classification following the standardized AFRICOVER system for the Africa region (Gregorio, 2005) was followed. It was used to classify the images into 7 main land cover/land use categories of: water (WAT), forest (FOR), agricultural land (AGR), woodland and grasslands (WG), settlement/urban (SET), dark bare soils (DAR), bright bare soils (BAR) (Table 2). Two classes of burnt scar (BSCAR), and cloud and cloud shadows (CL) were added in some scenes to enhance accuracy. The supervised classification allows the user to define areas of interest (AOIs) that identify and recognize features on the image.

**Table 2.** The land cover classification scheme used in the study.

| Generalized land cover classes | Description |
|---|---|
| 1. Buildup lands (settlements/urban) | All residential, commercial and industrial area, transportation infrastructure, settlements (may include green-house plastic covers). |
| 2. Water | Permanent open water, lakes, reservoirs, streams, dams, bays and estuaries. |
| 3. Forest | Deciduous forest land, evergreen forest land, mixed forest land. |
| 4. Agriculture | Cultivated large-scale farms, fallow, cultivated crops such as wheat, maize, beans, vegetables, coffee *etc.* |
| 5. Woodlands/grasslands | Dwarf shrubs and shrub land, wooded wetlands, grasslands. |
| 6. Bare bright soils | Very bright barren soils with barely any vegetation, rocky mountain. |
| 7. Dark soils | Composed of rocky mountains areas with little to no cover, wet and dark swampy areas where the pixel signature appears mixed. |
| 8. Cloud and cloud shadow | Clouds and cloud shadows present in some images. |
| 9. Burn scar | Burnt cover area in (Laikipia study area only). |

The AOI's (training samples) were collected for the various land cover/use categories, based on a) prior knowledge of the area, and b) uniformity in appearance using ArcGIS 10. The AOI's were used to create a signature file for each Landsat scene. The signatures were evaluated using histograms, band by band scatter plots and a separability analysis to obtain the expected error, covariance between bands in the classification for various feature combinations (Landgrebe, 2003). High covariance between bands is an indication of poor separability between cover categories, and therefore a need to refine class signatures to attain less covariance (ArcGIS_resources). The collection of signatures was repeated where the results were not satisfactory. The quality of the AOI's was analyzed using histograms, band scatter plots, and statistics reports derived from the training sample evaluation tool. Signatures were edited where separability results of the different cover classes was unsatisfactory. The signature file is used in classifying the entire scene. A maximum likelihood decision rule was used as a parametric rule (Richards and Jia, 2005) in the classification.

### 3.1.4. Accuracy assessment

Accuracy assessment analysis was carried out to validate the classification results. Reference points for the two study regions were generated aligned to the extent of the study area shapefile. A total of 300 points were generated for Laikipia and 125 points for the Athi River study area. The points were converted to kml files, for overlay on Google Earth imagery. This was necessary to assess the accuracy of classification results and modification of land cover categories. Historical Google Earth imagery (dating to January, 1976) was used as reference/validation (Zhao et al., 2014) for the 1984 cover categories, while recent Google Earth imagery (2010) was used to verify classification results for 2009 and 2010.

Frequency tabulations of classification results alongside referenced ground cover classes from Google Earth were developed. These were converted to pivot tables from where error/confusion matrixes detailing the overall accuracy, user accuracy and producer accuracy were calculated. The error matrix provides the users' accuracy and producer accuracy (Campbell, 2002) for the cover classes based on referenced points and classified output. The user accuracy present the number of correctly identified pixels in one class divided by the total number of sets recognized in that class.

The producer accuracy is obtained by dividing the number of correct pixels in one class by the total number of pixels derived from reference data. The USGS proposes a minimum accuracy level of 85 percent for LULC mapping using Landsat data (Anderson et al., 1976). A Kappa coefficient commonly used to measure map and classification agreement (Hudson and Ramm, 1987) was obtained. Bakr et al. (2010) indicate that the overall classification accuracy and Kappa statistics from the error matrix need to be close to 100% as much as possible.

### 3.1.5. Post processing and change detection

A degree of noise is often encountered in classification results due to isolated pixels or small regions of pixels in the image output. Post processing of the output raster reduces such noise, thereby improving the quality of classified output. Using ArcGIS tools, a majority filter (4*4) was applied which replaces cells in a raster based on the majority of their contiguous neighboring cells. Smoothing the class boundaries and clumping of the classified output was done. The final output was a land cover raster, which was converted to a polygon feature class from where the LULC area was calculated.

Detecting the amount and location of LULC changes is important particularly when seeking to understand drivers of change and the interaction of processes (Congalton and Green, 2009, Mundia and Aniya, 2006 and Yeh et al., 1996). Indeed, different methods to detect changes in LULC from remote images are available (Jensen, 2005 and Kashaigili and Majaliwa, 2010), and vary in complexity depending on the purpose of the studies. In this work two approaches, image differencing using the atmospherically corrected data and post classification comparison were employed. The image differencing technique is straight forward, easy to interpret and robust. The first date image is subtracted from the second date image, pixel by pixel, however, the result, is an absolute difference percentage value spread between 0% and 1%, with the brightest tone representing highly changed pixels. This does not provide detailed change information.

For a reasonable interpretation of the image differencing result, the output image of brightness values was classified into 3 categories. The Natural breaks (jenks) standardized classification method (in ArcMap) was used to identify class breaks that best grouped similar brightness values, while maximizing the differences between the classes. The 3 categories represented: (a) percentages close to 1 indicating areas of high change (red), (b) mid percentages indicating areas of moderate change (yellow) and, (c) percentages close to zero indicating areas of little to no change (white). The output map was then used in conjunction with the supervised classification land cover maps to visualize areas of changes in pixel values and correspondence to the post classification.

Post classification comparison approach was used to assess changes between the multi-date multispectral images. It is a common approach in change detection studies (Coppin et al., 2004, Kashaigili and Majaliwa, 2010 and Kiage et al., 2007), suitable in that it bypasses the challenges associated with the analysis of images acquired at different dates, by different sensors and at different times of the years (Yuan et al., 2005). However, the optimal accuracy of resulting LULC changes is highly dependent on individual classification which can be subject to errors (Zhang et al., 2002). Even so, changes in land cover classes are identified by comparing the multi date classified images, pixel by pixel using a change detection matrix. A change detection matrix table was derived by examining coverage (in sq. km) of individual classes in 1984, 2009 and 2010 (for Athi River).

### 3.1.6. Assessing rate of cover change

To assess dynamics of LULC changes in the study areas between 1984 and 2009/2010, for the different coverage classes the approach by Kashaigili and Majaliwa (2010) was applied. The average percentage change in cover was obtained by subtracting the area in 1984 (1st date) from coverage area in 2009/2010 (2nd date). The annual rate of change was computed by dividing the change between 1st date and the second date by number of years between the dates. The average percentage annual rate of change was derived by dividing the change between dates by the product of area of cover at 1st date multiplied by number of years, expressed as a percentage.

### 3.1.7. Population data and analysis

Population increase is associated with increased land use and cover conversion within watersheds (Lathrop and Haag, 2007). To evaluate the trend in population growth within the sub-watersheds comprising the study areas, the census data for 1989, 1999 and 2009 were utilized. The purpose of this step was to obtain an overall picture of trends in population growth, but whether these trends were a statistically significant driver of changes in the region is not tested. The census data follow administrative boundaries (i.e., districts and sub-locations), where the sub-locations are administrative units at a lower level than the districts and locations. The census population statistics in Kenya were obtained at this level, and aggregated to the district level. The 2009 census data obtained from Kenya National Bureau of Statistics (KNBS, 2009) were joined to the district layer and population numbers for each sub-watershed extracted using ArcGIS®, and the population density derived by dividing the total population per sub-watershed by area of the sub-watershed.

# 4. RESULTS AND DISCUSSION

## 4.1. Land cover maps and classification accuracy

The resulting maps indicate broad variations in LULC that occurred during the 25 years of study for Laikipia and Athi River. Figs. 2 and 3A–C show the land cover maps for 1984 and 2009/2010 for Laikipia and Athi River study regions. The overall accuracies of between the LULC maps were above 85%, and Kappa statistics were ranged between 0.78%

and 0.84% indicating a strong to moderate acceptable agreement between the classification map and the Google Earth reference information (Anderson et al., 1976).

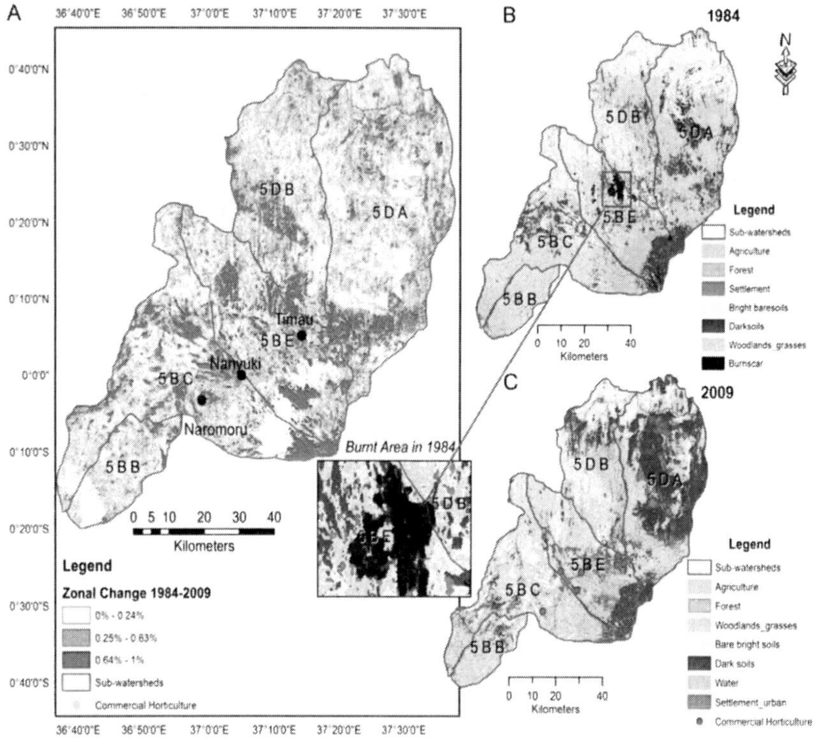


**Figure 2.** The maps illustrate the land use and land cover (LULC) types for Laikipia between 1984 and 2009. (A) Shows a zonal change detection map obtained by image differencing for Laikipia region (B) Map indicates the LULC in 1984 (C) Map indicates LULC in 2009. *Inset:* The burnt area in 1984, which was completely transformed by 2009.

## 4.2. Classification accuracy matrix

The overall accuracies of the LULC maps were above 80%, and Kappa statistics between 0.78% and 0.84% (Table 3) indicating a strong to moderate acceptable agreement between the classification map and the

Google Earth reference information (Anderson et al., 1976). The results of the overall classification accuracy and Kappa statistics were within acceptable classification requirements for the Landsat data. In Athi River, the producer accuracy tend to overestimate the WG cover, and the settlement (100%). It could be a likely result of misclassification of greenhouses as settlement and not area under agriculture. This is interesting because, plants are planted under the greenhouse structures. The water class seems to be under estimated in 1984 classification in both areas. This may be due to error in obtaining accurate training samples for the water cover. The producer accuracy for bright soils and dark soils were moderate, indicating slight misclassification most likely due to inability to distinguish bare and dark land in the Google imagery reference data. Moreover, fallow agricultural lands with high organic matter may appear dark, and so are damp areas near wetlands. Misclassification (Lu et al., 2004) of cover types while using Maximum Likelihood Classification (MLC) in remote sensing data is a common challenge often rectified by integration other geo-rectified data such as previous classified results, transportation networks, digital elevation models (Jensen, 1996 and Jensen, 2005) to correct the misclassifications.

## 4.3. Changes in cover area in Laikipia region

In both time periods the Laikipia study area (6852.5 sq. km) was predominantly covered by woody grassland (WG), bare bright soils (BAR), and dark barren soils (DAR), with proportions varying between 14.5% and 37.5% (Table 4). Forest (FOR) cover, agriculture (AGR), and settlement/urban occupied lesser proportions of the total area cover (Table 4).

Results show dynamic changes in cover proportions that occurred between 1984 and 2009. For example, the area under agriculture (AGR) increased from 496.3 sq. km (7.2%) to 659.1 sq. km (9.6%) at an average annual rate of 6.5 sq. km/year. WG which occupied vast cover area in both time periods (i.e. 37.5% and 34.3% respectively) declined by −222.1 sq. km, indicating an annual loss of −8.9 sq. km/year. BAR had the highest decline in cover (−675.4 sq. km), losing about −27 sq. km of land area annually (−1.4%). DAR increased from 994.2 to 1457.6 sq. km

(7% ), which was attributed to fallow agricultural lands during the off season agriculture period in the region. Besides, the prolonged 2008–2010 drought and low rainfall in region (Mwangi, 2012, Funk et al., 2003 and Pricope et al., 2013) may have had a severe impact in Laikipia, leading to the observed higher DAR cover. Droughts affect the terrestrial vegetation, reducing available cover types to bare grounds, and therefore the observed increase in DAR.

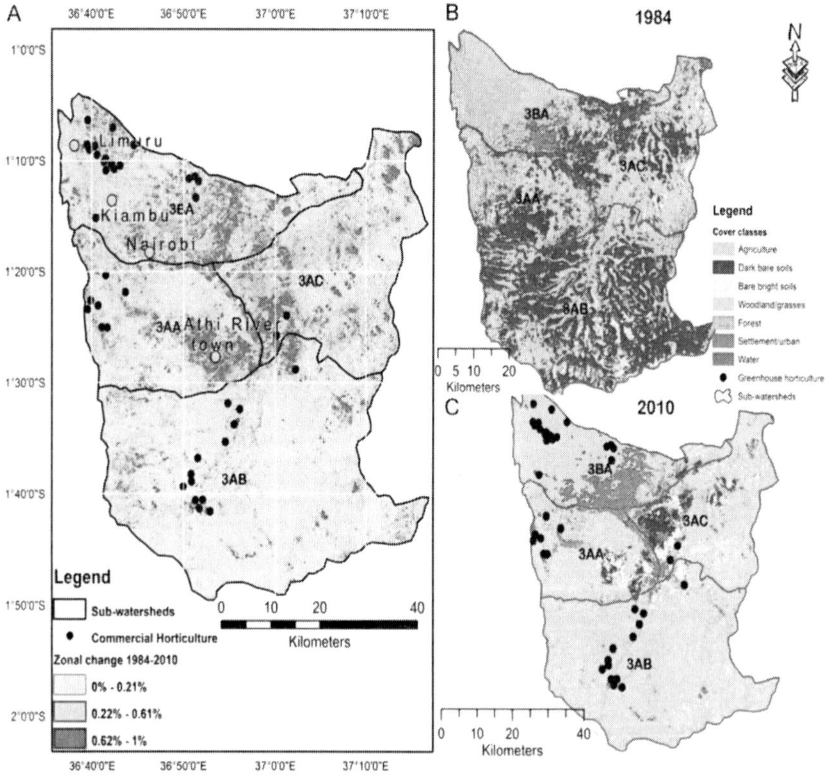


**Figure 3.** The maps illustrate the land use and land cover (LULC) types for Athi River region between 1984 and 2010. (A) Shows a zonal change detection map obtained by image differencing for Laikipia region, (B) Map indicates the LULC in 1984, (C) Map indicates the LULC in 2010.

**Table 3.** The table shows the accuracy assessment statistics including overall accuracy, producer accuracy, User accuracy and the kappa statistics for all cover types for the dates (1984, 2009 and 2010). *GE=Google Earth reference points; OA=Overall Accuracy; K=Kappa statistics.

| Classified land cover/land use | Athi River study area | | | | Laikipia study area | | | |
|---|---|---|---|---|---|---|---|---|
| | 1984 | | 2010 | | 1984 | | 2009 | |
| | User accuracy | Producer accuracy | User accuracy | Producer accuracy | User accuracy | Producer accuracy | User accuracy | Producer accuracy |
| Agriculture | 93.2 | 95.8 | 89.6 | 88.8 | 97.6 | 95.3 | 97.0 | 86.9 |
| Dark soils | 85.3 | 85.3 | 82.0 | 77.8 | 75.8 | 92.9 | 78.3 | 77.8 |
| Bright soils | 95.0 | 93.8 | 81.2 | 75.6 | 64.3 | 75.5 | 85.5 | 75.6 |
| Woodlands/grass | 79.8 | 100.0 | 87.3 | 96.5 | 80.6 | 87.5 | 85.7 | 93.3 |
| Forest | 96.9 | 81.6 | 85.5 | 86.0 | 83.5 | 76.7 | 76.7 | 89.3 |
| Settlement | 81.0 | 100 | 95.7 | 92.3 | 87.5 | 86.8 | 85.7 | 75.0 |
| Water | 73.4 | 83.3 | 86.6 | 89.9 | 87.0 | 62.0 | 86.7 | 90.0 |
| GE points | 125 | | | | 300 | | | |
| OA | OA=88% | | OA=89% | | OA=87% | | OA=85% | |
| K statistics | K=0.84 | | K=0.82 | | K=0.80 | | K=0.78 | |

**Table 4.** Coverage (sq. km) for each LULC category in 1984 and in 2009, changed area, and the average percentage annual rate of change during the study period in Laikipia. The average annual rate of change (change in area/time), and the % annual rate of change are represented as *Avrg.* ARC and % ARC respectively.

| Cover class | 1984 | | 2009 | | Change (sq. km) | *Avrg.* ARC | % ARC change |
|---|---|---|---|---|---|---|---|
| | Cover (sq. km) | % Cover (sq. km) | Cover (sq. km) | % Cover (sq. km) | | | |
| AGR | 496.3 | 7.2% | 659.1 | 9.6% | 162.8 | 6.5 | 1.3% |
| WG | 2573.1 | 37.5% | 2351.0 | 34.3% | −222.1 | −8.9 | −0.3% |
| BAR | 1936.2 | 28.3% | 1260.9 | 18.4% | −675.4 | −27.0 | −1.4% |
| DAR | 994.2 | 14.5% | 1457.6 | 21.3% | 463.4 | 18.5 | 1.9% |
| FOR | 699.4 | 10.2% | 574.9 | 8.4% | −124.5 | −5.0 | −0.7% |
| SET | 85.1 | 1.2% | 513.8 | 7.5% | 428.7 | 17.1 | 20.2% |
| BSCAR | 68.2 | 1.0% | 0.0 | 0.0% | −68.2 | −2.7 | −4.0% |
| WAT | 0.0 | 0.0% | 2.5 | 0.0% | 2.5 | 0.0 | 0.0% |
| CLD | 0.0 | 0.0% | 33.0 | 0.5% | 33.0 | 1.3 | 0.0% |
| **Total area (sq. km)** | 6852.5 | 100.0% | 6852.7 | 100.0% | | | |

*Note:* AGR = agriculture; WG = woodland and grasslands; BAR = bare bright soils; FOR = forest; DAR = dark soils and wet areas; SET = settlement/urban; BSCAR = burn scar; WAT = water; CLD = cloud and cloud shadow.

The burn scar in the 1984 imagery (68.2 sq. km) changed to other cover types by 2009 (Fig. 2). The results also indicate a decline in forest cover (−124.5 sq. km) during the study period, at an annual rate of −5 sq. km. This change was relatively minimal considering changes in the other main cover types in the area. Mount Kenya forest is the primary forest in this study area, and strongly influence the micro climate of the region. It was declared a biodiversity hotspot following rapid deforestation in the early 1980s, which perhaps explain the observed minimal rate of cover decline. Even so, controlled farming occurs on the forest edges, with plots allocated to farmers who tend to tree nurseries under the afforestation program.

The area under water increased from merely 0.001 sq. km to 2.5 sq. km. This estimate of the first classification seems low. However, but from a prior knowledge of the region, majority of surface waters are seasonal rivers, which often dry out to form "laggas" in dry seasons. In addition, unlike the water stored in dams for farm usage, and that occurring in ponds (visible in the scenes in Athi River), flowing rivers have high plant cover along the banks, making their accurate classification challenging. The increase is attributed to increased water abstraction into ponds/dams for irrigation water and not necessary an increase in surface waters. The rapid rise of commercial horticulture increase in the region (Justus and Yu, 2014 and Ulrich, 2014), prompted an increase in surface water abstractions points as reported in (Aeschbacher et al., 2005 and Kiteme and Gikonyo, 2002).

Interesting, the area under settlement/urban use (Table 4) increased intensely by 428.7 sq. km, at an annual rate of 17.1 sq. km (20.2%). This was by far the highest increase in cover/use area in the region for the last 25 years. However, this was not unexpected given the migration of rural populace in search of employment and land for development goals.

Considering the average percentage annual rate of change for the different LULC categories, BAR, FOR and WG declined annually at −1.4%, 0.7% and −0.3% respectively, while SET, DAR and AGR increased annually at 20.2%, 1.9%, and 1.3% respectively.

Change detection results via image differencing approach (Fig. 2 and Fig. 3A) indicated varied spatial changes in pixel values. The regions with a deep red color indicate areas where the change in pixel value varied between 0.64% and 100% (100% being a total change in pixel value) between 1984 and 2009/2010. The values between 0.26% and 0.63% indicated areas of moderate change, while 0–0.25% represented regions where little to no change in pixel value occurred. High change was noticeable in sub-watersheds 5BE and 5BC, while the rest of highly changed areas are along the border lines of the forest where wheat plantations extend into former forested area. Moreover, areas experiencing a decline in WG and the re-growth on the burn scar in 1984 classification image are shown as red or yellow color transitions where the change was not very high.

## 4.4. Changes in cover area in Athi River region

In 1984, the Athi River study region was primarily covered by woody grasslands (WG) and dark bare soils (DAR) occupying 1699.0 sq. km (40.5%) and 1538.2 sq. km (36.6%) respectively (Table 5). These proportions changed tremendously by 2010, where WG increased to 2611.6 sq. km (62.2%) and the DAR declined from 36.6% to 3.8%. The change in DAR cover (−1378.7 sq. km) indicated an average annual percentage decline of −3.4% during the 26 year period. The area under agriculture (AGR) increased from just 1.8% to 18.9% (Table 5), at an annual rate of change of 27.7 sq. km. Based on the results, AGR had the highest percentage annual rate of change (37.3%), followed by the area under water (17.2%). The increase in AGR area concurrent to the increase in area under water was most perhaps a result of increased dams and water storage ponds for irrigation agriculture, and open sewage reservoir near the city. Clusters of horticultural greenhouses are observed in the upper part of the study area towards Limuru in Kiambu district, and the lower southern parts (Fig. 1B). The clusters were not present in the 1984 image, but developed at later dates. Unlike rain-fed agriculture, intensive horticulture utilizes all-year round irrigation systems for crop production. Furthermore, the expansion of the Nairobi city and the sub-urban (UNEP, 2002) impelled the need for food supplies prompting a rapid conversion of cover to agricultural land. In fact, the results show significant increase in settlement/urban use from 85.4 sq. km to 272.8 sq. km (8.4%). Dark bare soils (DAR) and forest cover (FOR) declined annually by 53 sq. km and 21.9 sq. km respectively.

**Table 5.** Land cover area, changed area and the rate of change over 26 years (1984–2010) in Athi river region. The average annual rate of change (change in area/time), and the % annual rate of change are represented as *Avrg.* ARC and % ARC respectively.

| Cover class | 1984 | | 2010 | | Change (sq. km) | *Avrg.* ARC | % ARC |
|---|---|---|---|---|---|---|---|
| | Cover (sq. km) | % Cover (sq. km) | Cover (sq. km) | % Cover (sq. km) | | | |
| AGRI | 74.2 | 1.8% | 794.6 | 18.9% | 720.3 | 27.7 | 37.3% |
| SET | 85.4 | 2.0% | 272.8 | 6.5% | 187.4 | 7.2 | 8.4% |
| BAR | 126.8 | 3.0% | 256.1 | 6.1% | 129.3 | 5.0 | 3.9% |
| WG | 1699.0 | 40.5% | 2611.6 | 62.2% | 912.5 | 35.1 | 2.1% |
| DAR | 1538.2 | 36.6% | 159.5 | 3.8% | − 1378.7 | − 53.0 | − 3.4% |
| FOR | 668.4 | 15.9% | 97.9 | 2.3% | − 570.5 | − 21.9 | − 3.3% |
| WAT | 1.21 | 0.03% | 6.62 | 0.2% | 5.4 | 0.2 | 17.2% |
| CL | 5.83 | 0.1% | 0.00 | 0.0% | − 5.8 | | |
| **Total Area (sq. km)** | 4199.1 | 100.0% | 4199.1 | 100.0% | | | |

*Note:* AGR=agriculture; SET=settlement/urban; BAR=bare bright soils; WG=woodland/grasslands; DAR=dark soils and wet areas; FOR=forest; WAT=water; CL=cloud and cloud shadow.

Considering the two study areas, the observed percentage annual rate of change between cover classes in Athi River were higher, and with a smaller total land area (4199.1 sq. km) compared to Laikipia which has a total land area of 6852.0 sq. km. The increase in LULC was in the order of AGR<WAT<SET<BAR<WG (37.7%, 17.2%, 8.4%, 3.9< and 2.9% respectively) while DAR<FOR declined at −3.4% and −3.3%.

The change detection results (Fig. 3A) illustrate major changes in pixel values (deep red color) around Athi river town and Nairobi city, patterns consistent with the expansion of settlement/urban use in the classification result of 2010 (Fig. 3). Besides increasing economic activity around the city, the reported spread of greenhouse horticulture in lower zones of the study area (Kithiia, 2012b) could partially explain the detected high to moderate changes in pixel values, implying changes in LULC. Most farm workers commute between Athi river town center and the farm locations, which explain the concentrated changes around the town. Moderate to high change in pixel values occurred around Limuru, Kiambu, as indicated by the yellow to red colorations, identifying wider regions of cover transformation likely result of urban sprawl and expanding cultivation.

## 4.5. Land covers transformation

### 4.5.1. Laikipia

The change detection matrix (Table 6) highlights changes that occurred between LULC classes during the 25 years of study. The bold diagonal numbers indicate the cover area (sq. km) that remained unchanged between 1984 and 2009, while the other numbers show the area of specific LULC (sq. km) that changed to another class. The magnitude of change from one class to another varied between the two dates.

**Table 6.** Change detection matrix for different LULC categories (in sq. km) between 1984 and 2009 for Laikipia.

| Cover in 1984 (sq. km) | Cover in 2009 (sq. km) | | | | | | | | | |
|---|---|---|---|---|---|---|---|---|---|---|
| | AGR | WG | BAR | DAR | FOR | SET | BSCAR | WAT | CL | Total |
| AGR | **31.1** | 238.3 | 87.9 | 44.2 | 32.3 | 61.2 | 0.0 | 0.1 | 1.1 | 496.1 |
| WG | 250.1 | **1270.2** | 284.9 | 477.8 | 89.8 | 197.6 | 0.0 | 0.1 | 2.6 | 2573.1 |
| BAR | 150.9 | 299.9 | **816.9** | 355.1 | 2.0 | 309.8 | 0.0 | 0.1 | 1.1 | 1935.8 |
| DAR | 52.9 | 317.8 | 40.1 | **529.0** | 28.5 | 20.6 | 0.0 | 0.1 | 5.3 | 994.2 |
| FOR | 16.4 | 185.0 | 1.2 | 23.3 | **419.3** | 30.9 | 0.0 | 2.2 | 21.1 | 699.4 |
| SET | 9.9 | 13.3 | 25.1 | 6.0 | 1.0 | **29.0** | 0.0 | 0.0 | 0.8 | 85.1 |
| BSCAR | 2.5 | 25.7 | 5.3 | 22.4 | 1.9 | 9.5 | **0.0** | 0.0 | 1.0 | 68.2 |
| WAT | 0.0 | 0.0 | 0.0 | 0.0 | 0.0 | 0.0 | 0.0 | **0.0** | 0.0 | 0.01 |
| CL | 0.0 | 0.0 | 0.0 | 0.0 | 0.0 | 0.0 | 0.0 | 0.0 | **0.0** | 0.00 |
| Total | 513.7 | 2350.2 | 1261.4 | 1457.8 | 574.81 | 658.6 | 0.0 | 2. 54 | 32.8 | 6852.1 |

*Note:* AGR=agriculture; SET=settlement/urban; BSAR=burn scar; BAR=bare bright soils; WG=woodland/grasslands; DAR=dark soils and wet areas; FOR=forest; WAT=water; CL=cloud and cloud shadow. The diagonal bold numbers indicate cover areas that remained unchanged between 1984 and 2009.

For example, along the row of woodland and grasslands (WG), 250.1 sq. km of WG cover were converted to AGR, 1270.2 sq. km remained unchanged, 284.9 sq. km was changed to BAR. At same time, 477.8 sq. km changed to DAR, 89.8 sq. km changed to FOR, 197.9 sq. km changed to SET, 0.1 sq. km changed to water and 2.6 sq. km was cloud cover and shadows.

The entire burn scar area (BSCAR 100%) changed to WG and DAR at 25.7 and 22.4 sq. km respectively, by 2009 (Table 6). Forest fires arising from human activities or natural causes leave burnt scars on the landscape. The scar is spectrally different from the bare soils. It was interesting to observe its transformation over time. More importantly, monitoring the type of cover that may succeed fire episodes could provide useful information to resource and conservation managers. Fire episodes are common in savannas, and essentially promote re-generation of new vegetation cover promoting life.

Large area under FOR remained unchanged (419.3 sq. km),( Tables 6, and 7) which as explained earlier, is attributed to international recognition of the region as a rich harbor for threatened reservoirs of plant and animal life (UNEP, 2008) and therefore increased surveillance to monitor illegal deforestation and charcoal burning on the forest lines. Even so, some portions of forest cover changed to other LULC e.g. WG (185 sq. km), SET (30.9 sq. km) (Table 6).

A summary of cover classes that underwent change, or remained unchanged during the study period is shown in Table 7. In the table, the percentage unchanged cover indicates the percentage (%) area of the original area (1984s) which remained unchanged for the duration of study. Additionally, the percentage changed represent cover class area that changed to other classes. For instance, 94% of AGR cover changed to other uses, 51% of woodlands/grasslands, 58% of the bare bright soils *etc.* Interestingly, a small area remained under agriculture (31.1 sq. km) by 2009. This was unexpected given the increased large-scale greenhouse commercial horticulture in the region. However, given that the greenhouses have plastic covers, the pixel signature was interpreted as settlement or urban, and not agriculture use.

**Table 7.** The observed percentage (%) changes of respective cover classes in Laikipia between 1984 and 2009.

| Cover class | Unchanged (sq. Km) | Changed (sq. Km) | Cover unchanged (%) | Cover changed (%) |
|---|---|---|---|---|
| Agriculture | 31.1 | 455.1 | 6 | 94 |
| Woodlands and grasslands | 1270.2 | 1302.9 | 49 | 51 |
| Bare bright soils | 816.9 | 1118.9 | 42 | 58 |
| Dark bare soils | 529.0 | 465.3 | 53 | 47 |
| Forest | 419.3 | 280.1 | 60 | 40 |
| Settlement/urban | 29.0 | 56.2 | 34 | 66 |
| Burn scar | 0.0 | 68.2 | 0 | 100 |

Rain-fed agriculture is highly dependent on ample rainfall, and therefore rainfall fluctuations and unpredictability could limit agricultural yield and therefore a decline in agricultural area. This explanation suffices for the observed moderate percentage change in bare bright soils (58%) and woody grasslands (51%). Low or decline in vegetative cover exposes soils with varying degree of mixed backgrounds.

### 4.5.2. Athi River region

The change detection matrix for Athi River is shown in Table 8. The bold numbers on the diagonal illustrate the cover area which remained unchanged between the two years, while the other numbers (sq. km) illustrate the changes to other cover/use classes. LULC change between classes was highly dynamic, with widespread cover conversion. For instance, along the row of WG 179.6 sq. km was changed to agriculture (AGRI), 112.6 sq. km to settlement/urban (SET), 187.6 sq. km to bare bright soils (BAR), 1157.4 sq. km remained unchanged, 50.6 sq. km to dark bare soils (DAR), 8.6 sq. km to forest (FOR), and 2.6 sq. km was covered by water (WAT).

The results in Table 9 present a summary of changed and unchanged cover classes between 1984 and 2010 in the Athi River study area. The percentage cover unchanged illustrate the percentage of a certain cover class that remained unchanged, while percentage changed represent the percentage cover area that changed to other land cover/use between the study dates. The area originally under agriculture underwent the least change (13.3%), while 1433.4 sq. km (93.2%) of DAR cover changed to other LULC classes, primarily WG (1223.8 sq. km) as shown in Table 8. There were tremendous changes in forest cover (87.9%) with 587.4 sq. km converted to other uses (Table 9). From the LULC cover map (Fig. 3.), the expansion of Nairobi city and growth of suburbs towns, industrial centers and mining companies on the periphery of the city enhanced LULC transformation. Reports indicate that towns of Athi River and Mlolongo town (in sub-watershed 3AB), have emerged as strong grounds for horticulture farming and industrial growth in recent years (UNEP, 2008), a finding corroborated by Justus and Yu (2014). Expansion of intensive horticulture into arid and semi-arid marginal areas has implications for rangelands managements and resource use.

**Table 8.** Change detection matrix for different LULC categories (in sq. km) between 1984 and 2010 for Athi River.

| Cover in 1984 (sq. Km) | Cover in 2010 (sq. km) | | | | | | | | |
|---|---|---|---|---|---|---|---|---|---|
| | AGRI | SET | BAR | WG | DAR | FOR | WAT | CL | Total |
| AGRI | **64.4** | 3.3 | 0.1 | 2.8 | 0.0 | 3.5 | 0.0 | 0.0 | 74.2 |
| SET | 13.6 | **42.6** | 5.1 | 19.1 | 2.2 | 2.3 | 0.5 | 0.0 | 85.4 |
| BAR | 3.7 | 0.6 | **15.7** | 106.5 | 0.1 | 0.1 | 0.2 | 0.0 | 126.8 |
| WG | 179.6 | 112.6 | 187.6 | **1157.4** | 50.6 | 8.6 | 2.6 | 0.0 | 1699.0 |
| DAR | 102.3 | 60.6 | 42.1 | 1223.8 | **104.8** | 2.3 | 2.4 | 0.0 | 1538.2 |
| FOR | 429.9 | 52.3 | 4.8 | 99.3 | 0.3 | **81.0** | 0.7 | 0.0 | 668.4 |
| WAT | 0.6 | 0.2 | 0.0 | 0.1 | 0.0 | 0.1 | **0.2** | 0.0 | 1.2 |
| CL | 0.5 | 0.6 | 0.7 | 2.5 | 1.5 | 0.0 | 0.0 | 0.0 | 5.8 |
| Total Area | 794.6 | 272.8 | 256.1 | 2611.6 | 159.5 | 97.9 | 6.6 | 0.0 | 4199.1 |

*Note*: AGR=agriculture; SET=settlement/urban; BAR=bare bright soils; FOR=forest; CL=cloud-cloud shadow. The diagonal bold numbers indicate unchanged cover 1984 and 2010.

**Table 9.** Observed percentage changes of respective cover classes in Athi River between 1984 and 2010.

| Cover class | Unchanged (sq. Km) | Changed (sq. Km) | % Cover unchanged | % Cover changed |
|---|---|---|---|---|
| Agriculture | 64.4 | 9.9 | 87% | 13.3% |
| Settlement/ urban | 42.6 | 42.8 | 50% | 50.1% |
| Bare bright soils | 15.7 | 111.1 | 12% | 87.6% |
| Woodlands/ grasslands | 1157.4 | 541.6 | 68% | 31.9% |
| Dark bare soils | 104.8 | 1433.4 | 7% | 93.2% |
| Forest | 81.0 | 587.4 | 12% | 87.9% |
| Water | 0.2 | 1.0 | 16% | 83.9% |

### 4.5.3. Population density and LULC dynamics

This section examines the classification results in conjunction with population trends in the sub-watersheds (Fig. 4 and Fig. 5). In both study areas, 1999 had the highest number of people per sq. km in the sub-watersheds, but this declined in 2009. In fact, the decline was lower than the 1989 trend for the Laikipia sub-watersheds except 5BB, which borders highly populous areas along foot zones of Aberdare ranges.

The rapid increase in population density in 1999 is attributed to the influx of people into the sub-watersheds following the expansion of, and economic vibrancy brought about by the regional horticulture farming. This increase in population density of the sub-watersheds, plausibly explain the observed increase in area under settlement (20.2%) between 1984 and 2009 in Laikipia (Fig. 4B). This was exceptional in that all other average annual percentage rates of cover change were within a range of 1–2%, except the burn scar that entirely transformed. Recent decline (2009 is the latest census in Kenya) in population density in both study area could imply movement of people to other areas. It is also possible that the 2008 post-election violence affected demographic structure in these regions, pushing “immigrants” out, as a result of ethnic-political issues/instigations. Reports indicate that the horticulture sub-sector suffered huge loss following the violence, due to poor employee turn outs at work places (Ksoll et al., 2009 and Ksoll et al., 2010).

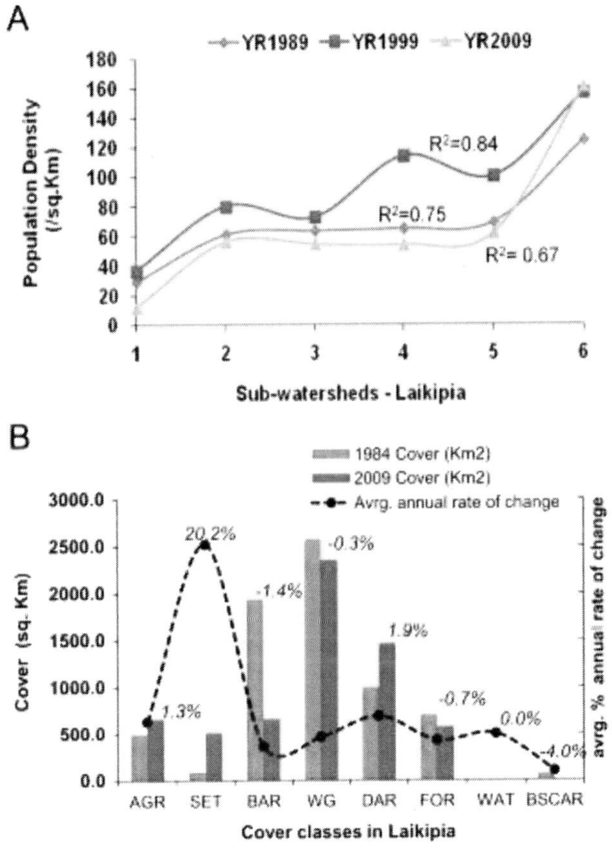


**Figure 4.** (A) The plots illustrate the population trends (1989, 1999, and 2009) in the sub-watersheds (upper panel) and, (B) LULC dynamics (lower panel) in Laikipia study area.

In Fig. 5A, sub-watershed 3BA (which is shared by Nairobi and Kiambu counties in Fig. 3) has had high population density through time. The rapid urban growth of Nairobi city during the last three decades has played a big role in land transformation due to influx of migrating population in search of employment and economic gains. Following the implementation of the economic reforms and open-door policy after independence (1963), there was increased immigration of people from rural to urban areas. The effect of increased population density (Fig. 5A and B) for example in sub-watershed 3BA enhanced LULC transformation, associated with characteristic settlement problems and environmental degradation (Olima, 2001). Emerging sub-urban towns in the outer skirts of the city, also rapidly influence LULC transformation. For instance, Athi River town was largely an industrial and mining center, but has greatly transformed following influx of large-scale horticulture companies. Altogether, unlike Laikipia where settlement/urban use increased parallel to increase in population density (1989–1999), in Athi River, agriculture use had a remarkable annual increase (37.3%), followed by area under water (17.2%) and settlement/ urban (8.4%).

Increased demand for agricultural food commodities by the expanding city population reasonably explains the observed spike in the average percentage annual increase in land under agriculture and water. Intensive small-holder farming and greenhouse horticulture farming are common in Limuru and other parts of Kiambu districts, forming a key source for agricultural produce. Good road networks within the cities, to the Jomo Kenyatta International Airport *etc.* provide means to deliver produce to target markets. These changes have affected other cover classes; FOR and DAR, which have declined at −3.3 and −3.4% respectively.

## 4.5.4. Variations in detected changes and limitations

Classifying cover in the highly heterogeneous landscape of very bright to very dark surfaces, mosaics of low grass cover and interspaced with woody shrubs and agriculture presented an interesting situation. Using few LULC classes produced in accurate representation of cover types, however, this was overcome by increasing the number of cover classes and selecting separable signatures, providing more accurate land cover maps. These are important factors towards achieving satisfactory classification results.

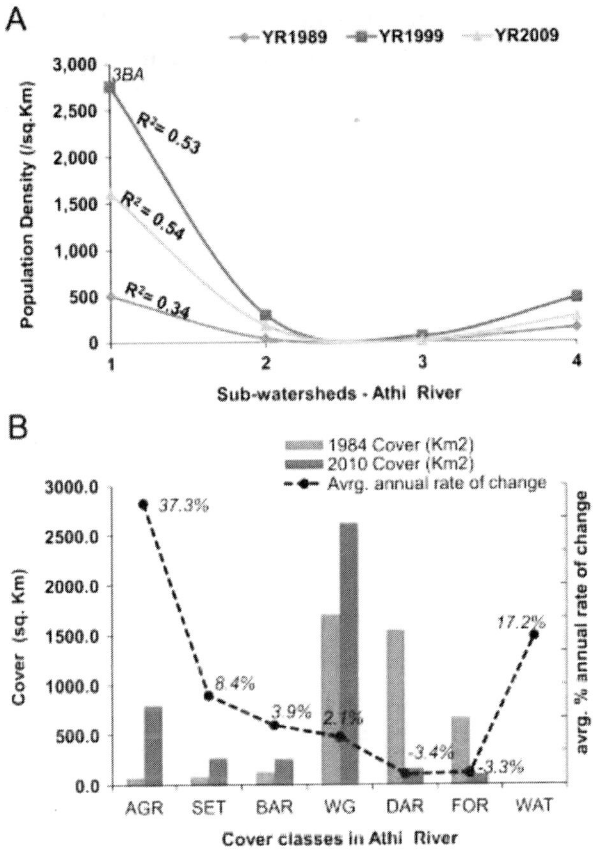


**Figure 5.** (A). The plots illustrate the population trends (1989, 1999, and 2009) in the sub-watersheds (upper panel) and, (B) LULC dynamics (lower panel) in Athi River study area.

Also, during classification some LULC categories were spectrally mixed making it difficult for complete separation following the supervised classification. For example, problems were encountered in classification of green landscapes within settlement/urban areas, with some pixels erroneously classified as agriculture use especially within Nairobi city, Athi River towns. The 2010 classification output for Athi River show a vast area under WG, which at first present the impression of over-estimation of the cover type. The acquisition date of the imagery coincided with drought seasons. The presence of high WG cover unlike the expectations may indicate regional differences in the intensity of the drought across the two regions. In addition, the Athi River basin has a high percent of grass cover mixed with low shrubs, typical of savanna lands (Kithiia and Ongwenyi, 1997). This can be confirmed from the grasslands data layer created by World Resource Institute (WRI) and used as overlay in Fig. 1B. Water coverage in Laikipia in 1984 appears underestimated, mainly due to poor signatures. Surface waters had substantial cover along banks which introduced mixed pixels and therefore misclassification as green corridors of agriculture/woodlands along the waterways. To improve the final LULC results, an ArcGIS overlay was applied and each cover class visually examined to counter check accuracy of classification, thereby increasing accuracy and the quality of the LULC maps (Shalaby and Tateishi, 2007).

### 4.5.5. Implications

The proportion of agriculturally productive land in both study regions is small compared to the vast arid and semi-arid land (ASALs) primarily covered by grasslands ecosystem. Grasslands are rich in biodiversity and concurrently represent fragile environments that are extremely vulnerable to habitat destruction (Hogan, 2012 and Shiyomi and Koizumi, 2001). The occurrence of grassland on easily farmed topography invite for monoculture production, livestock keeping *etc.* that transforms the landscape. As such the rapid decline of grassland ecosystems due to human activities is linked to dramatic loss in biodiversity and decline of keystone species ( Ceballos et al., 2010).

In the ASALs of sub-Saharan Africa, anthropogenic interference enhances effects of aridity and land degradation, causing high erosion rates and evapo-transpiration from bare land surfaces. This is a common problem that challenges resource management and conservation, and necessitate urgent tactics to improve the condition of grasslands (MEA, 2005).

In Kenya, conservation efforts to protect remaining patches of forests around the city of Nairobi for istance, Karura forest, Ngong forest, are manned through the ministry of Natural Environment Management Agency (NEMA). However, efforts are often complicated by ineffective policies and corruption issues. Besides, positive growth of the horticulture sub-sector is an important economic engine to the country, through improved Gross Domestic Product (GDP), foreign income from fresh produce exports and job creation.

Even so, it is important to evaluate LULC impelled by expansion of the sub-sector, and therefore seek sustainable approaches to achieve positive socio-economic growth while practicing stewardship of the natural environment and its resources. Egoh et al. (2011)shows that focusing conservation on areas that are important for provision of various ecosystem services could benefit biodiversity and vice versa, and more so if similar management regimes are needed. High population density within the sub-watersheds creates competition for food and land, which implies that both agricultural systems, and the remaining natural landscapes, are vulnerable to continued degradation, increasing biodiversity loss (Tilman et al., 2001) and dysfunctional ecosystems (Hector and Bagchi, 2007).

The study results are helpful for policy makers to better understand and address the complex relationships among LULC, population growth, urbanization, and environment. Considering the findings, important policy suggestions may include intensifying the regulation and administration of urban land, balancing cultivated land and forestry, and encouraging the development and utilization of unused land if environmental impact assessments permit. Comparable suggestions have been proposed for rapidly growing urban areas (Yang et al., 2012). Other studies (Ikiel et al., 2013) propose encouraging the use of unexploited land by local government to reduce environmental damage and relieve pressure on the land demand for urban use.

# 5. CONCLUSION

An investigation was carried out to determine the dynamics of LULC changes in sub watersheds within Laikipia and Athi regions, regions experiencing an increased growth of intensive commercial horticulture, and population growth. The study integrated various approaches to understand occurring LULC changes and likely environmental effects. The specific objectives were; to develop maps of LULC, quantify the dynamics and rate of LULC change and an evaluation of population trends in the sub-watersheds. Remote satellite data obtained from Landsat-5 (MSS, TM), Landsat-7 ETM+ sensors and census demographic data were utilized. The supervised classification results were assessed for accuracy using Google Earth archived images, and counter checked with prior land cover map, obtaining reasonable moderate to high accuracy levels in overall accuracy, user accuracy, producer accuracy and Kappa statistic. Post classification change detection enabled identification of localities of change, amount and type of change and the rate of change from one use to another between the periods of study.

Examining the results, both study areas have undergone notable changes in LULC in the 25 years of study. Agriculture and settlement/urban centers increased, and moreso in the Athi River basin, with land changes advancing towards more marginalized semi-arid areas of the sub-watersheds, that used to be pastoral grounds/ranches and wildlife habitats. Concurrently, forest cover declined tremendously and more-so in the upper north zones of the Athi River basin, a likely result of increased intensive horticulture and population expansion. Expansion of Nairobi city and the connecting infrastructure development is good for economic growth. For horticulture growers, the city provides a closer and ready market for the produce, and also faster access to cargo freights to international markets. However, inability to balance the social – economics with the environmental needs makes such expansion chaotic and unsustainable.

From an environmental management stand point, the rapid and vast transformation of land within the fragile unstable arid to semi-arid (ASALs) sub-watersheds in the study area is unsettling. This is because LULC changes in such areas pose increased vulnerability to climate variations and related impacts on resources, people etc. ASALs are dominated by grasslands and woodlands, which support a high diversity of wildlife, promoting local and international tourism and therefore the need for better management. Conserving forested areas, controlling encroachment into fragile landscapes and monitoring rates of LULC transformation are necessary precautions towards improving overall health of the watersheds. This understanding will aid development of policy and regulatory measures that address the challenges in watershed resource use and management. Efficient management strategies allow delivery of multiple ecosystem services, (e.g. through better land use policies), while conserve threatened biodiversity (Mt. Kenya and the grasslands ecosystems). This is important and very crucial given increasing human induced pressures on the highly vulnerable semi arid and arid ecosystems in the region.

## REFERENCES


1. Abd El-Kawy, O.R., Rød, J.K., Ismail, H.A., Suliman, A.S., 2011. Land use and land cover change detection in the western Nile delta of Egypt using remote sensing data. Appl. Geogr. 31 (2), 483–494.

2. Adger, W.N., Arnell, N.W., Tompkins, E.L., 2005. Adapting to climate change: perspectives across scales. Glob. Environ. Chang. 15 (2), 75–76.

3. Aeschbacher, J., Liniger, H.P., Weingartner, R., 2005. River water shortage in a highland-lowland system: a case study of the impacts of water abstraction in the Mount Kenya region. Mt. Res. Dev. 25, 155–162.

4. Anderson, J.R., Hardy, E.E., Roach, J.T., Witmer, R.E., 1976. Land Use and Land Cover Classification System for Use With Remote Sensor Data. Geological Survey, Washington, DC.U.S.964.

5. ArcGIS_resources, 2016. An overview of the Image Classification toolbar. ⟨http://resources.arcgis.com/en/help/main/10.11/index.html//00nv00000008000000⟩.

6. Bajocco, S., De Angelis, A., Salvati, L., 2012. A satellite-based green index as a proxy for vegetation cover quality in a Mediterranean region. Ecol. Indic. 23 (0), 578–587.

7. Bakr, N., Weindorf, D.C., Bahnassy, M.H., Marei, S.M., El-Badawi, M.M., 2010. Monitoring land cover changes in a newly reclaimed area of Egypt using multi-temporal Landsat data. Appl. Geogr. 30 (4), 592–605.

8. Baldyga, T.J., Miller, S.N., Driese, K.L., Gichaba, C.M., 2008. Assessing land cover change in Kenya's Mau Forest region using remotely sensed data. Afr. J. Ecol. 46 (1), 46–54.

9. Barrow, C.J., 2006. Environmental Management for Sustainable Development. Routledge, London (UK), New York (NY).

10. Becht, R., Odada, E.O., Higgins, S., 2005. Lake naivasha: experience and lessons learned bried lake basin management initiative. Managing lakes and basins for sustainable use, a report for lake basin managers and stakeholders. ILEC, Kusatsu, pp. 277–298.

11. Berger, P., 1989. Rainfall and agro-climatology of the Laikipia plateau, Kenya. African Studies Series A7. Institute of Geography, University of Berne, Switzerland.

12. Campbell, J.B., 2002. In: Introduction to Remote Sensing, 3rd ed. Guilford Press, New York.

13. Ceballos, G., Davidson, A., List, R., Pacheco, J., Manzano-Fischer, P., Georgina, S.B., Juan, C. 2010. Rapid Decline of a Grassland System and Its Ecological and Conservation Implications. PLoS ONE 5(1): e8562.

14. Congalton, R.G., Green, K., 2009. Assessing the Accuracy of Remotely Sensed Data: Principles and Practices. CRC Press, Taylor and Francis. Coppin, P., Jonckheere, I., Nackaerts, K., Muys, B., Lambin, E., 2004. Review article digital change detection methods in ecosystem monitoring:a review. Int. J. Remote Sens. 25, 1565–1596.

15. Decurtins, S., 1992. Hydro-geological investigations in the Mt. Kenya subcatchment of the Ewaso Ng'iro River (African Study Series A6). Geographica Bernensia, University of Bern, Switzerland.

16. De Leeuw, P.N., Grandin, B.E., Solomon, B., 1991. Introduction to the Kenyan rangelands and Kajiado district. In: Solomon, B., De Leeuw, P.N., Grandin, B.E., Neate, P.J.H. (Eds.), Maasai herding: An analysis of the livestock production system of Maasai pastoralists in eastern Kajiado district, Kenya, International Livestock Center for Africa (ILCA), Ethiopia.

17. Egoh, B.N., Reyers, B., Rouget, M., Richardson, D.M., 2011. Identifying priority areas for ecosystem service management in South African grasslands. J. Environ. Manag. 92 (6), 1642–1650.

18. Francis, K.M., 2014. Biodiversity in Fragmented Key Resource Areas: Persistence of Plant Diversity And Productivity in An Intensively Grazed Key Resource Area (ITC Dissertation no. 256). University of Twente, Netherlands.

19. Funk, C., Michaelson, J., Verdin, J., Artan, G., Husak, G., Senay, G., 2003. The collaborative historical African rainfall model: description and evaluation. Int. J. Climatol. 23, 47–66.

20. Gichuki, F.N., Liniger, H.P., MacMillan, L.C., Schwilch, G., Gikonyo, J.K., 1998. Scarce water: exploring resource availability, use and improved management Resources, actors and policies — towards sustainable regional development in the highland – lowland system of Mount Kenya. East South Afr. J. 8, 15–28. Special Issue.

21. Giri, C., Zhu, Z., Reed, B., 2005. A comparative analysis of the Global Land Cover 2000 and MODIS land cover data sets. Remote Sens. Environ. 94 (1), 123–132.

22. Gregorio, A.D., 2005. Land cover classification system (LCCS), version 2: classification concepts and user manual. In: Proceedings of the FAO Environment and Natural Resources Service Series, No. 8. FAO, Rome. Global Facility for Disaster Reduction and Recovery (GFDRR), 2012. Kenya post-disaster needs assessment (PDNA) 2008–2011 drought. Available at ⟨http://www.gfdrr.org/sites/ gfdrr/files/Kenya_PDNA_Final.pdf⟩ (accessed 11.09.15.).

23. Google Earth Imagery, 2010. DigitalGlobe-2010. ⟨http://www. earth. google. com⟩.

24. Haberl, H., Erb, K.H., Krausmann, F., Gaube, V., Bondeau, A., Plutzar, C., Gingrich, S., Lucht, W., Fischer-Kowalski, M., 2007. Quantifying and mapping the human appropriation of net primary production in earth's terrestrial ecosystems. Proc. Natl. Acad. Sci. USA 104 (31), 12942.

25. Hector, A., Bagchi, R., 2007. Biodiversity and ecosystem multifunctionality. Nature 448 (7150), 188–190.

26. Hogan, C., 2012. Habitat destruction. ⟨http://www.eoearth.org/view/article/153224⟩.

27. Hudson, W.D., Ramm, C.V., 1987. Correct formulation of the Kappa coeffi- cient of agreement. Photogramm. Eng. Remote Sens. 53 (4), 421–422.

28. Huho, J.M., Mugalavai, E.M., 2010. The effects of droughts on food security in Kenya. Int. J. Clim. Chang. : Impacts Responses 2, 61–72.

29. Ikiel, C., Ustaoglu, B., Dutucu, A.A., Kilic, D.E., 2013. Remote sensing and GIS-based integrated analysis of land cover change in Duzce plain and its surroundings (north western Turkey). Environ. Monit. Assess. 185 (2), 1699–1709.

30. Imbernon, J., 1999. Pattern and development of land use changes in Kenyan highland since 1950s. Agric. Ecosyst. Environ. 76, 67–73.

31. Jensen, J.R., 1996. Introductory Digital Image Processing: A Remote Sensing Perspective, 2nd Ed. Prentice-Hall Inc, New Jersey, U.S.

32. Jensen, J.R., 2005. Information extraction using artificial intelligence. In: Keith, C.C. (Ed.), Introductory Digital Image Processing – A Remote Sensing Perspective 3rd ed. , Prentice Hall Series in Geographic Information Science, Saddle River, NJ, USA, pp. 407–429. Sect. D.

33. Justus, F., Yu, D., 2014. Spatial distribution of greenhouse commercial horticulture in Kenya and the role of demographic,

infrastructure and topo-edaphic factors. ISPRS Int. J. Geo-Inf. 3 (1), 274–296.

34. Kashaigili, J.J., Majaliwa, A.M., 2010. Integrated assessment of land use and cover changes in the Malagarasi river catchment in Tanzania. Phys. Chem. Earth Parts A/B/C 35 (13–14), 730–741.

35. Kasperson, J.X., Kasperson, R.E., Turner, B.L., 1996. Regions at risk: exploring environmental criticality. Environ.: Sci. Policy Sustain. Dev. 38 (10), 4–29.

36. Kaufman, Y.J., 1994. The atmospheric effect on the separability of field classes measured from satellites. Remote Sens. Environ. 18 (1), 21–34.

37. Kiage, L.M., Liu, K.B., Walker, N.D., Lam, N., Huh, O.K., 2007. Recent land- cover/use change associated with land degradation in the Lake Baringo catchment, Kenya, East Africa: evidence from Landsat TM and ETM. Int. J. Remote Sens. 28 (19), 4285–4309.

38. Kirui, K.B., Kairo, J.G., Bosire, J., Viergever, K.M., Rudra, S., Huxham, M., Briers, R.A., 2013. Mapping of mangrove forest land cover change along the Kenya coastline using Landsat imagery. Ocean Coast. Manag. 83 (0), 19–24.

39. Kiteme, B.P., Gikonyo, J., 2002. Preventing and resolving water use con- flicts in the Mount Kenya highland-lowland system through water users associations. Mt. Res. Dev. 22 (4), 332–337.

40. Kithiia, S.M., 2012a. Effects of sediments loads on water quality within the Nairobi River Basins, Kenya. Int. J. Environ. Prot. 2 (6).

41. Kithiia, S.M., 2012b. Water Quality Degradation Trends in Kenya over the Last Decade. In: Voudouris, Kostas, Voutsa, Dimitra (Eds.), Water Quality Monitoring and Assessment, InTech

42. Kithiia, S.M., Ongwenyi, S.O., 1997. Some problems of water quality degradation in the Nairobi River sub-basins in Kenya. In: Proceedings of the Rabat Symposium S4 on Freshwater

Contamination, April 1997, 243. IAHS Press, Wallingford, UK, pp. 121–127.

43. KNBS, 1989. National population and housing census results 1989: population distribution by age, sex, and administrative units, Nairobi, Kenya. Available online: ⟨https://opendata.go.ke⟩ (accessed on 25 October 2014).

44. KNBS, 1999. National population and housing census results 1999: population distribution by age, sex, and administrative units, Nairobi, Kenya. Available online: ⟨https://opendata.go.ke⟩ (accessed on 25 October 2012).

45. KNBS, 2009. National population and housing census results 2009: population distribution by age, sex, and administrative units, Nairobi, Kenya. Available online: ⟨https://opendata.go.ke⟩ (accessed 25. 10. 14.).

46. Ksoll, C., Macchiavello, R., Morjaria, A., 2009. Guns and roses: the impact of the Kenyan post election violence on flower exporting firms. CSAE Working Paper Series. Centre for the Study of African Economies, University of Oxford.

47. Ksoll, C., Macchiavello, R., Morjaria, A., 2010. The effect of ethnic violence on an export-oriented industry. BREAD Working Paper. Bureau for Research and Economic Analysis of Development.

48. Landgrebe, D.A., 2003. In: Signal Theory Methods in Multispectral Remote SensingJohn Wiley and Sons, Hoboken, New Jersey508.

49. Lathrop, R.G., Haag, S.M., 2007. Assessment of Land Use Change and Riparian Zone Status in the Barnegat Bay and Little Egg Harbor Watershed: 1995–2002–2006. ⟨http://bbp.ocean.edu/Reports/Riparian %20Zone%20Assessment.pdf⟩.

50. Liniger, H., Gichuki, F.N., Kirochi, G., Njeru, L., 1998. Pressure on the land: the search for sustainable use in a highly diverse environment. East South Afr. Geogr. J. (ESAG), 8. Special Number.

51. Liu, Y., Huang, X., Yang, H., Zhong, T., 2014. Environmental effects of landuse/cover change caused by urbanization and policies in Southwest China Karst area – A case study of Guiyang. Habitat Int. 44 (0), 339–348.

52. Liverman, D., Emilio, F.M., Ronald, R.R., Paul, C.S., 1998. In: People and Pixels: Linking Remote Sensing and Social ScienceNational Academy of Sciences, Washington, D.C.

53. Lu, D., Mausel, P., Brondízio, E., Moran, E., 2004. Change detection techniques. Int. J. Remote Sens. 25 (12), 2365–2401.

54. Lu, D., Xu, X., Tian, H., Moran, E., Zhao, M., Running, S., 2010. The Effects of Urbanization on net primary productivity in Southeastern China. Environ. Manag. 46 (3), 404–410.

55. Mahmood, R., Pielke Sr, R.A., Hubbard, K.G., 2006. Land use/land cover change and its impacts on climate. Glob. Planet. Chang 54 (1–2).

56. Maingi, J.K., Marsh, S.E., 2001. Assessment of environmental impacts of river basin development on the riverine forests of eastern Kenya using multi-temporal satellite data. Int. J. Remote Sens. 22 (14), 2701–2729.

57. Mwangi, E., 2012. Droughts in the GHA: a case study of the 2010/2011 drought in Kenya. Available at ⟨http://www.gfdrr.org/sites/gfdrr/files/ Kenya_PDNA_Final.pdf⟩.

58. MDC, 1993. Matrix Development Consultants. Nairobi Informal Settlements: An Inventory. USAID/DISC, Arlington, USA.

59. MEA, 2005. Millennium Ecosystem Assessment Synthesis Report. Island Press, Washington, D.C (accessed 12.11.14.). http://www.maweb.org/ en/Synthesis.aspx.

60. MoPND, 1994. Ministry of Planning and National Development. Kiambu District Development Plan (1994–96). Ministry of Planning and National Development, Government Printer, Nairobi, Kenya.

61. Mundia, C.N., Aniya, M., 2006. Dynamics of landuse/cover changes and degradation of Nairobi City, Kenya. Land Degrad. Dev. 17 (1), 97–108

62. Mungai, D.N., Ong, C.K., Kiteme, B., Elkaduwa, W., Sakthivadivel, R., 2004. Lessons from two long-term hydrological studies in Kenya and Sri Lanka. Agric. Ecosyst. Environ. 104 (1), 135–143.

63. Muriuki, G., Seabrook, L., McAlpine, C., Jacobson, C., Price, B., Baxter, G., 2011. Land cover change under unplanned human settlements: a study of the Chyulu Hills squatters, Kenya. Landsc. Urban Plan. 99 (2), 154–165.

64. Mutiga, J.K., Shadrack, T.M., Su, Z., Tsehaie, W., Robert, B., 2010. Water allocation as a planning tool to minimise water use conflicts in the Upper Ewaso Ng'iro North Basin, Kenya. Water Resour. Manag. 24, 3939–3959.

65. Neubert, M., Meinel, G., 2005. Atmospheric and Terrain correction of IKONOS Imagery Using ATCOR3. Leibniz Institute of Ecological and Regional Development (IOER), Dresden, Germany.

66. Notter, B., 2003. Rainfall-Runoff Modeling of Meso-Scale Catchments in the Upper Ewaso Ng'iro Basin, Kenya. University of Berne, Bern, Switzerland.

67. Notter, B., MacMillan, L., Viviroli, D., Weingartner, R., Liniger, H., 2007. Impacts of environmental change on water resources in the Mt. Kenya region. J. Hydrol. 343 (3–4), 266–278.

68. Oehl, F., Sieverding, E., Ineichen, K., Mäder, P., Boller, T., Wiemken, A., 2003. Impact of land use intensity on the species diversity of arbuscular mycorrhizal fungi in agroecosystems of Central Europe. Appl. Environ. Microbiol. 69 (5), 2816–2824.

69. Olima, W.H.A., 2001. The dynamics and implications of sustaining urban spatial segregation in Kenya: experiences from Nairobi metropolis. In: Proceedings of the Conference Paper, Lincoln Institute of Land Policy, Cambridge, MA, USA, July 25–28, 2001.

70. Owiti, G.E., Oswe, I.A., 2007. Human impact on lake ecosystems: the case of Lake Naivasha, Kenya. Afr. J. Aquat. Sci. 32, 1.

71. Patz, J.A., Olson, S.H., 2008. Land use/Land change and Health. Int. Encycl. Public Health, 15–21.

72. Pricope, N.G., Husak, G., Lopez-Carr, D., Funk, C., Michaelsen, J., 2013. The climate-population nexus in the East African Horn: emerging degradation trends in rangeland and pastoral livelihood zones. Glob. Environ. Change 23 (6), 1525–1541.

73. Rattray, J.M., 1960. The Grass Cover of Africa. 49. FAO (Food and Agriculture Organization of the United Nations), Rome, Italy168 FAO Agriculture Studies.

74. Richards, J.A., Jia, X., 2005. In: Remote Sensing Digital Image Analysis – An IntroductionSpringer, Berlin.

75. Richter, R., 1996. Atmospheric correction of satellite data with haze removal including a haze/clear transition region. Comput. Geosci. 22 (6), 675–681.

76. Schneider, L.C., Gil Pontius Jr, R., 2001. Modeling land-use change in the Ipswich watershed, Massachusetts, USA. Agric. Ecosyst. Environ. 85 (1–3), 83–94.

77. Shalaby, A., Tateishi, R., 2007. Remote sensing and GIS for mapping and monitoring land cover and land-use changes in the Northwestern coastal zone of Egypt. Appl. Geogr. 27 (1), 28–41.

78. Shisanya, C.A., 1990. The 1983–1984 drought in Kenya. J. East. Afric. Res. Dev. 20, 127–148.

79. Shiyomi, M., Koizumi, H., 2001. Structure and function in agroecosystem design and management (Advances in Agroecology). CRC press, Boca Raton FL.435.

80. Smith, J.A., Tzeu, L.L., Ranson, K.J., 1980. The lambertian assumption and landsat data. Photogramm. Eng. Remote. Sens. 46 (9), 1183–1189.

81. Thomas, M.K., Liniger, H.P., 1994. Development of a streamflow model for rural catchments in Kenya. Laikipia—Mt. Kenya Papers (Discussion paper-11, Nanyuki).

82. Tibaijuka, A., 2007. Nairobi city development strategy top priority for 21st Century future of the Kenyan capital. In: Barr, J., Shisanya, C. (Eds.), Nairobi and its Environment, United Nations Environment Programme, Nairobi.

83. Tiffen, M., Mortimore, M., 1992. Environment, population growth and productivity in Kenya: a case study of Machakos district. Dev. Policy Rev. 10 (4), 359–387.

84. Tilman, D., Fargione, J., Wolff, B., D'Antonio, C., Dobson, A., Howarth, R., 2001. Forecasting agriculturally driven global environmental change. Science 292 (281–284).

85. Townshend, J.R.G., 1994. Global data sets for land applications from the advanced very high resolution radiometer: an introduction. Int. J. Remote Sens. 15 (17), 3319–3332.

86. Turner II, B.L., Lambin, E.F., Reenberg, A., 2007. The emergence of land change science for global environmental change and sustainability. Proc. Natl. Acad. Sci. 104 (52), 20666–20671

87. Ulrich, A., 2014. Export-oriented horticultural production in laikipia, kenya: assessing the implications for rural livelihoods. Sustainability 6 (1), 336–347.

88. UNEP, 2002. African Environment Outlook: GEO-4. United Nations Environment Programme, Nairobi.

89. UNEP, 2008. Africa: Atlas of our Changing Environment. Division of Early Warning and Assessment (DEWA). United Nations Environment Programme, Nairobi, Kenya.

90. Weismann, U., 1998. Sustainable Regional Development in Rural Africa: Conceptual Framework and Case Studies from Kenya. African Studies Series, A14. Geographica Bernesia, Berne Switzerland.

91. West, P.C., Narisma, G.T., Barford, C.C., Kucharik, C.J., Foley, J.A., 2010. An alternative approach for quantifying climate regulation by ecosystems. Front. Ecol. Environ. 9 (2), 126–133.

92. Wulder, M.A., White, J.C., Goward, S.N., Masek, J.G., Irons, J.R., Herold, M., 2008. Landsat continuity: issues and opportunities for land cover monitoring. Remote Sens. Environ. 112 (3), 955–969.

93. Yang, H., Xie, P., Ni, L., Flower, R.J., 2012. Pollution in the Yangtze. Science 337 (6093), 410-a.

94. Yang, J., Gong, P., Fu, R., Zhang, M.H., Chen, J., Liang, S., 2013. The role of satellite remote sensing in climate change studies. Nat. Clim. Chang 10, 875–883.

95. Yeh, A., Gar, A., Xia, L., 1996. Urban growth management in Pearl River Delta: an integrated remote sensing and GIS approach. ITC J. 1 77–78.

96. Yuan, F., Sawaya, K.E., Loeffelholz, B.C., Bauer, M.E., 2005. Land cover classification and change analysis of the Twin Cities (Minnesota) Metropolitan Area by multitemporal Landsat remote sensing. Remote Sens. Environ. 98 (2–3), 317–328.

97. Zhang, Q., Wang, J., Peng, X., Gong, P., Shi, P., 2002. Urban built-up land change detection with road density and spectral information from multi-temporal Landsat TM data. Int. J. Remote Sens. 23 (15), 3057–3078.

98. Zhao, Y., Gong, P., Yu, L., Hu, L., Li, X., Li, C., Wang, X., 2014. Towards a common validation sample set for global land-cover mapping. Int. J. Remote Sens. 35 (13), 4795–4814.

# CHAPTER 11

# Managing the Selenium Content in Soils in Semiarid Environments through the Recycling of Organic Matter

**R. Garcia Moreno,[1] R. Burdock,[2] María Cruz Díaz Álvarez,[3] and J. W. Crawford[2]**

[1] Faculty of Sciences, University of La Coruña, Zapateira, 15001 A Coruña, Spain

2 Judith and David Coffey Chair, Faculty of Agriculture, Food and Natural Resources, University of Sydney, Suite 411 Biomedical Building, 1 Central Avenue, Australian Technology Park, Eveleigh, Sydney, NSW 2015, Australia

3 CEIGRAM (Centre for Studies and Research on Agricultural and Environmental Risk Management), School of Agricultural Engineering, Polytechnic University of Madrid, Ciudad Universitaria S/N, 28040 Madrid, Spain

## ABSTRACT

Around 30% of the world's population suffers from either a lack of one or more essential micronutrients, or the overconsumption of these minerals, which causes toxicity. Selenium (Se) is a particularly important micronutrient component of the diet with a well-documented and wide-ranging role in maintaining health. However, this important micronutrient can be lacking because soil and crop management are focused on high yields to the detriment of the quality of crops required to ensure a healthy human diet. Currently around 15% of the global population has selenium deficiency. This paper focuses on Se availability in semiarid soils and how micronutrients can be effectively managed through the recycling of organic matter. Because many mineral reserves are being exploited unsustainably, we review the advantages of using organic by-products for the management of the biofortification of Se in crops. This type of practice is particularly useful in arid and semiarid environments because organic matter acts as a reservoir for Se, preventing bioaccumulation and leaching. There are also potential local economic benefits from using organic by-products, such as manures and sewage sludge.

## 1. INTRODUCTION

At least 60% of the world's population either lacks one or more essential mineral elements or consumes food containing high amounts of toxic mineral elements [1]. Mineral malnutrition is a widespread problem in both developing and developed countries. This situation is particularly serious for some micronutrients, such as Fe, Zn, I, Se, Ca, Mg, and Cu [2, 3]. In the specific case of Se, 15% of the world's population is already Se deficient [2].

Gupta et al. [4] stated that, in addition to the lack of studies on the ability of plants to uptake minerals, there are insufficient analyses of soil that determine the total nutrient contents. Similarly, there are no studies of the impact of different soil management practices on the concentration and distribution of micronutrient concentrations in the different edible parts of crop plants.

The micronutrient status of a plant can be measured from the leaves because leaves contain the highest amounts of micronutrients. Micronutrient deficiency is easily detected in younger leaves, whereas toxicity can be detected in later stages of development in older leaves [5]. Several factors control the lack of Se content in plants including the genetic variety, soil management, soil type, and climate. The lack of micronutrient content in plants is common in humid temperate and tropical regions due to intense soil leaching caused by the high number of rain events.

Ekholm et al. [6] studied the trends in the mineral and trace element contents in fruits, vegetables, and cereals in Finland over the last 30 years. They found that the content of most minerals has decreased in all crops. Interestingly, the only exception is Se, the content of which has increased as a result of the use of selenium-supplemented inorganic fertilisers over the last 20 years. Tennant et al. [3] found the same pattern of decline in UK crops and noted that analogous trends have occurred in different countries that share very similar historical farming management strategies, based mainly on the adoption of modern genetic varieties of crops and agronomic practices related to the acceleration of the growth rates of plants. These practices include growth in higher temperatures, increased light intensity, increased $CO_2$ concentrations, and higher irrigation rates.

Micronutrient cycling in soils is closely related to the organic matter content and crop residue management [7]. Therefore, soil and crop mismanagement that leads to the loss of organic matter leads to a reduction in the mineral content of soils and consequently the mineral content of plants. Biofortification can be achieved either through the use of genetically improved crops for mineral uptake into the edible parts of the plant or through the use of specific fertilisers that increase the phytoavailability of some minerals to specific crops [2]. However, inorganic fertilisers are increasingly expensive to manufacture, distribute, and apply. These fertilizers also have environmental impacts, including an increase in greenhouse gas emission, the unsustainable exploitation of mineral resources, and the mineral enrichment of the environment leading to soil contamination. Biofortification is mainly dependent on the specific chemical forms of micronutrients in soils and the subsequent uptake by plants [8]. To reinforce this strategy, it is necessary to know the forms of the mineral elements that are available to plants, as well as the limitations and the phytoavailability of these elements in the rhizospheres [9]. In the specific case of selenium, the plant roots will obtain this element through the uptake of selenate, selenite, or organoselenium compounds by the plant roots [3].

Agronomic biofortification strategies to increase the mineral concentrations in the edible parts of plants generally rely on the application of mineral fertilisers and the improvement of the solubilisation and mobilisation of soil micronutrients. In the case of selenium, the application of soil and foliar Se fertilisers has been shown to have beneficial effects on animal health and human nutrition. Soil applications are suggested for cropping under late-season moisture and heat stress [3, 5, 6].

The consumption of many key minerals is unsustainable according to Hueso et al. [10], with current rates of consumption leading to the depletion of global mineral reserves in the next 50 years. According to the US Geological Survey [10], almost all of the selenium produced worldwide is currently recovered from anode slimes during Cu electrolysis and to a lesser extent during Ni and Zn electrolysis. According to this same source, the global Se reserves are 172,000 t Se, which is based on Cu deposits. Thus, if one-third of the world's arable land (500 Mha) is fertilised with 10 g Se/ha, 5000 t Se/yr will be consumed, and the global Se reserve would be depleted in less than 40 years. Thus, the use of minerals as fertilisers has to be prioritised to ensure that the nutritional demands are achieved or other strategies must be implemented to avoid the exhaustion of mineral reserves. To these ends it is also important to obtain more information on the geographical distribution of micronutrients in relation to nutrient sampling in crops and the relationship with epidemiological studies to evaluate the relationship between animal and human diseases and the geochemical environment and soil and plant management related to nutrition crops [5].

## 2. IMPORTANCE OF SELENIUM IN THE HUMAN DIET

Se was discovered in 1817 by Berzelius; however, it was not until 140 years later that its nutritional requirement was determined [12]. The Se cycle is represented in Figure 1. Although Se is not essential for plant growth, this element has beneficial effects that promote the growth of plants and it is essential in a healthy diet for humans and other mammals [11]. In general, micronutrients, such as Se, are required as cofactors for enzymes or as part of the protein structure involved in DNA synthesis and repair, the prevention of oxidative damage to DNA, and the maintenance of the methylation of DNA [13]. Because of these functional roles, Se deficiency in mammals can result in several physiological disorders, such as cardiomyopathy (Keshan) and osteoarthritis diseases (Kashin-Beck), pancreatitis, asthma, inflammatory response syndrome, malfunctioning of immune system, decreased response to viral infections, and decreased fertility and thyroid functioning [13].

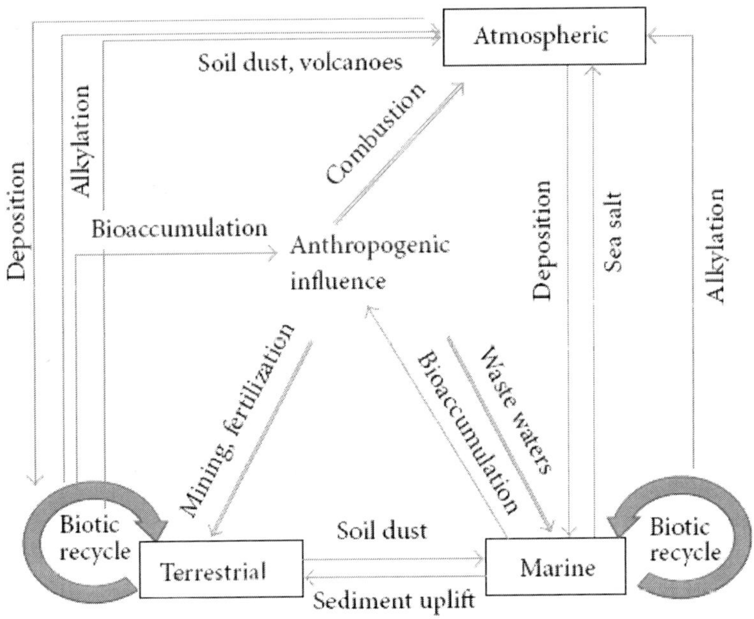


**Figure 1.** Global Se cycle considering the anthropogenic influence (extracted from [11]).

The individual human Se intake ranges from 3 to 7000 g Se/day worldwide, with most cases at the lower end of the range [14]. For beneficial effects, the US Government recently recommended dietary Se allowances of 55 and 70 g/day for women and men, respectively [12] and the current standards designated by the Department of Health (1991) in the UK set the upper safe limit to 400 g Se/day/person. Broadley et al. [15] noted that the Se intake in the UK has declined from 60 g Se/d to 29–39 g Se/d by the end of the 90s, and in other EU countries, the intakes are even lower at 30 and 38 g Se/d for females and males, respectively, based on the results of blood and serum testing.

Most of the researches conducted on Se and human health report a beneficial effect of this element in the prevention of cancer and cardiovascular disease [15–22]. Against this, Wu et al. [23] did not find any beneficial effects in human Australian males who consumed a supplementary Se diet with biofortified wheat (193 m/L). In these individuals, the beneficial effects were tested through biomarkers of cancer and cardiovascular diseases. However, as the authors noted, any beneficial effects may have been related to specific genotypes, and further studies are needed to confirm the conclusions. Se deficiency has been associated with liver, muscle, and heart disease in animals [24]. As mentioned, Se has one of the narrowest ranges in humans between dietary deficiency and toxic levels: less than 40 g $day^{-1}$ to more than 400 g $day^{-1}$ [25]. Thus, although global Se toxicity in humans is far less widespread than Se deficiency [26], it is still important to understand the biophysicochemical processes that regulate its bioavailability in nutrition [27]. Furthermore, Se is chemically similar to sulphur in plants and it is therefore metabolised through many of the same pathways, with many plants preferentially uptaking Se over S. Above a critical leaf dry matter concentration in the range of 10 to mg Se/kg plant dry weight, selenium toxicity begins to limit growth in most plants. Se toxicity in crops occurs mainly on seleniferous soils, and only plants that exhibit genetic tolerance to this type of soil can be successfully grown [2, 7].

The selenium level in most soils is generally less than 1 mg Se/kg soil; however, the selenium content in seleniferous soils can be as high as 4 to 100 mg Se/kg soil. The selenium content of plants in most soils is less than 1 mg/kg plant dry weight, whereas most plants grown in seleniferous soils show selenium levels in the range of 1 to 10 mg/kg plant dry weight. In the case of Se-hyperaccumulator plants, this can increase to between 1000 to 15,000 mg/kg [7]. These differences in concentration can be visualized in Figure 2, which shows potential areas where livestock may be at risk for selenium deficiency or toxicity in Australia. The major species of Se in plants include selenate, which is translocated directly from the soil and is less readily bound to soil components than selenite, and selenomethionine (SeMet) and selenocysteine (SelCys), both of which are biosynthesised by plants [15, 28]. The Se content in wheat and other cereals is in the form of SeMet with lower amounts of SelCys and selenate. Se-enriched crops exhibit a higher Se content than the corresponding natural plants; nevertheless, the treatment of crops must be performed with caution to reduce the risk of toxicity [29].

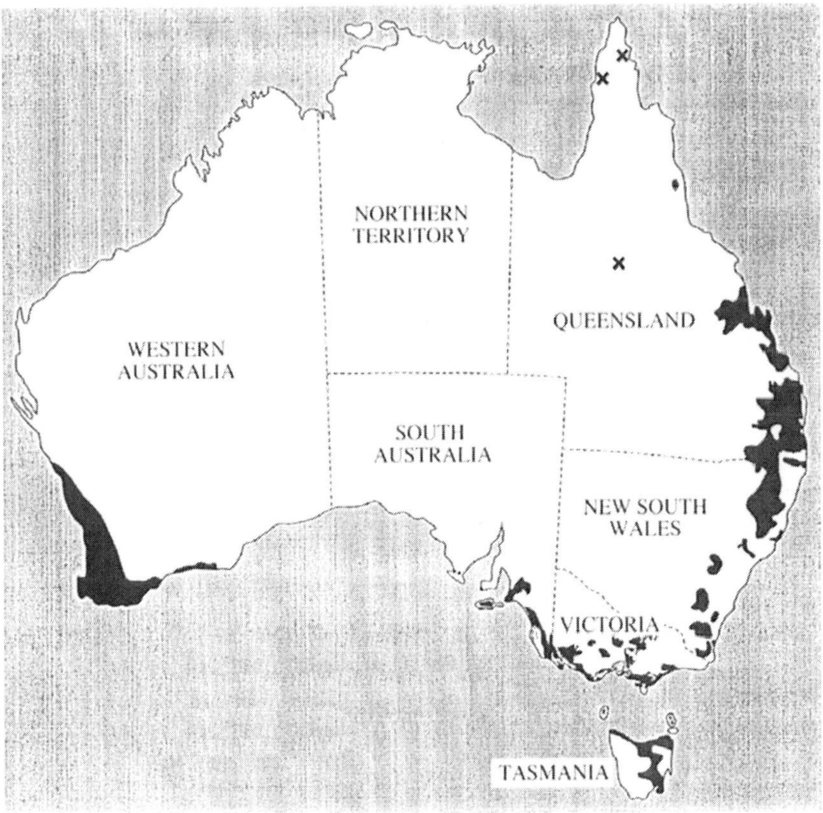


**Figure 2.** Map of Australia showing potential areas where livestock may be at risk for selenium deficiency or toxicity (extracted from [34]).

The absorption of Se in humans is not affected by the nutritional human status and approximately 80% of the absorbed selenium originates from food. The studies that have compared the bioavailability of different forms of Se in humans concluded that organic forms are more bioavailable than selenate and selenite [24, 30]. In fact Se-methyl-selenocysteine and its -glutamyl-derivative, which are found in a number of edible plants of theAllium and Brassica families, have been studied for their potent anticancer effects. These selenoproteins have very high antioxidant properties and are highly beneficial in counteracting diseases related to oxidative stress. As well as the benefits from selenoproteins, the Se obtained in the diet can be metabolised to small-molecular-weight species that are believed to exhibit antitumour effects [30].

At sufficiently high doses, the Se metabolites can also induce toxicity in animals and humans. The effects of the Se toxicity are highly dependent on the form of Se. Se compounds can easily form an anion that generates superoxide in the presence of thiols, such as glutathione, which results in redox cycling. Studies have shown that the toxic effects of Se are due to this oxidative-stress mechanism [31, 32]. In fact, superoxide has been shown to be generated from selenite and diselenides, such as selenocysteine, but not from selenite, in the presence of reduced glutathione [31, 32]. As a result of their inability to generate superoxide, SeMet nor Se-methyl-selenocysteine has a relatively low toxicity to cells in culture or to animals or humans. Nakamuro et al. [33] found that selenodiglutathione, which is an intermediate chemical species in the formation of superoxide from selenite and glutathione, is more toxic than selenite itself. Hasegawa et al. [34] found that another mechanism of Se toxicity includes the inhibition of Se methylation, which represents the major detoxification pathway for Se; the inhibition of this pathway results in the accumulation of selenide hepatotoxic compounds. Although studies on Se toxicity suggest that the organic forms may be more toxic than the inorganic forms after long-term consumption because the organic forms are more easily incorporated into tissues, there is no conclusive evidence that proves this hypothesis [34].

Finally, as noted by Rayman [29], it is difficult to define the optimal intake of Se at the individual level because its concentration in the body is dependent on a large number of factors, such as which functions of Se are most relevant for a certain stage of a disease, which species of Se is predominant in the Se source, the health conditions in the receptor, the adequacy of intake of other nutrients and their interferences, the presence of additional stressors, and the ability to produce selenoproteins.

## 3. THE ROLE OF SOIL FORTIFICATION

According to Levander and Burk [21], wheat (Triticum aestivum L.) represents the major dietary source of Se worldwide and therefore the decrease in Se intake has been related to changes in the Se content of cereal grains and to the decrease of Se in soils. This hypothesis is supported by decreases in intake and the trend over time in the UK to consume a larger proportion of grain grown in the UK that has a lower Se content compared with imported grain grown in North America, where the soils contain higher amounts of Se [15]. Zhu et al. [35] made a comparison of the Se content in rice and wheat between different countries and crops over the last 40 years, as shown in Figure 3.

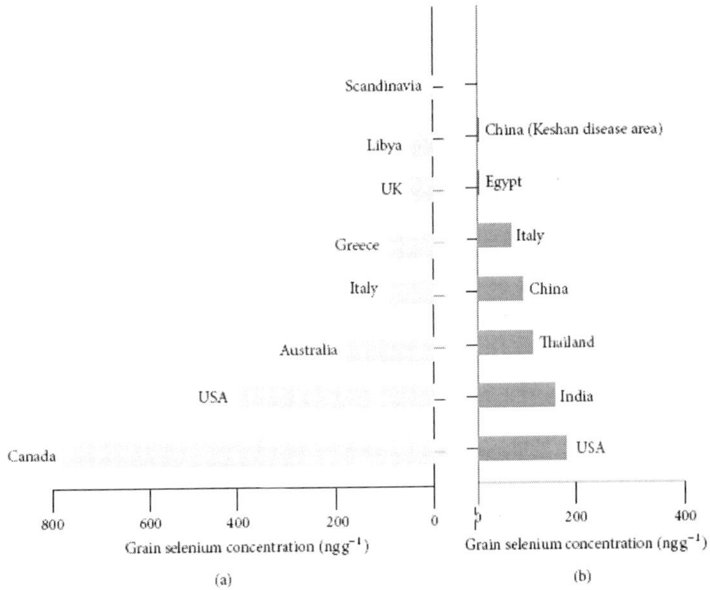


**Figure 3.** Comparison of Se levels in wheat (a) and rice grains (b) in different countries over the last 40 years (extracted from Zhu et al. [35]).

Lyons et al. [31] proved that the fertilisation of wheat crops in Australia with Se fertilisers is a cost-effective method for the improvement of the concentration of organic Se in grain, which results in an incremental uptake of Se in animal and human diets. As mentioned above, the important issue associated with the management of Se fertilisation is the control of the amount of Se added to avoid toxicity. The authors added Se in the form of selenate at fertilising rates of up to 120 g/ha in field trials and 500 g/ha in pilot trials in soil and added 330 g/ha of foliar Se applied with low S concentrations (2–5 mg S/kg soil). These researchers did not observe any symptoms of toxicity in the crop even if the application is 20 times higher that safest applications. A soil application of 10 g Se/ha is found to be safe for raising Se content of crops in Finland.

In the case of Se applied as selenite, they observed a critical amount of 70 mg/L required to cause growth inhibition, whereas selenate did not affect the crops in a solution concentration of 150 mg/L. They recommended using application rates of 200 g selenate/ha to avoid toxicity in wheat and to obtain tissue levels of Se below any toxic concentration. In later studies, Lyons et al. [36] found that the concentrations of Se in diverse germplasms did not vary in different cultivars and was in the range of 5 to 720 g/kg, regardless of the genotypic variation.

Finland has conducted fortification of crops since 1981 and has obtained excellent results: only the Se content exhibited an increasing trend in different crops compared to the decreasing trend in the concentrations of other micronutrients of up to 10-fold in most crops [6]. Eurola et al. [37] measured the Se content in 125 food items before and after fortification in Finland and found that the Se content had increased in all cases. The total intake was distributed across cereals (26%), meat (29%), dairy products (20%), eggs (10%), and fish (9%). In addition, only those diets with a very exceptional composition provide less than 0.05 mg or more that 0.2 mg of Se per 10 MJ (Mega Joules). In the case of cereals, of which wheat is the most important, their contribution to the total intake of Se almost tripled from 1971 to 1991. The biofortification of pastures and forages has been shown to increase the Se content in the diet by increasing the Se content in livestock [4, 5]. The last authors demonstrated that Se content in pastures and forages prevents disorders amongst grazing livestock, such as muscular dystrophy disorders.

Broadley et al. [38] used selenate fertiliser in wheat crops and found that the crop total recovery, including grain and straw, ranged from 20% to 35%. These researchers added up to 10 g Se/ha to the crop, and, if the straw is removed, approximately 6.5–8 g Se/ha is not recovered. For this reason, to ensure the long-term sustainability, the fate of Se in soils and in the food chain must be known before any biofortification strategy is widely implemented. The residual Se may be leached, volatilised, retained in the soil, strongly absorbed by iron or aluminium oxides, or retained in the soil in nonsoluble elemental forms.

Eurola et al. [37, 39] concluded that Se fertilisation significantly affects the Se content of cereals and other crops. However, the same authors also noted that the Se content varies considerably on an annual basis and between different locations mainly due to the specificity of the fertilising rates and the different soil and climatic conditions.

It is clear that there are individuals and populations that would benefit from a higher Se intake level. However, it is important to be aware of the specific geochemical environment and the baseline intake in any specific country or region because necessary additional intake in one region or country might be excessive in another [18]. Beside these health-related considerations, other elements that should be considered are the cost-effectiveness of adding Se to food sources, such as through biofortification or the recycling of organic by-products, and the environmental effects of the fertiliser and soil management strategies. For example, if the Se fertilisers are economically and environmentally expensive, other sources of recycling, as well as the management of different soils to conserve and retain the surplus of Se in soils to avoid their lost through leaching, evaporation, and absorption, must be considered.

# 4. SE MANAGEMENT FOR CROP FERTILISATION IN SEMIARID ENVIRONMENTAL CONDITIONS

In the case of semiarid and arid environments, the lack of micronutrients and macronutrients in crops is mainly due to high crop yield, insufficient return of crop residues to soil, low organic matter, and immobilisation of most micronutrients in the soil [3]. This is the situation in most Australian soils used to produce cereals [14], especially those used to produce cereals in the southern regions, which are mainly alkaline and particularly calcareous or sodic. These conditions drive chemical constraints for agricultural production associated with the alkaline soil environment and the dry land climate, which include macronutrient and micronutrient deficiencies and some toxicity related to high sodicity and salinity. More generally, alkaline soils represent an important proportion of world soils that sustain crop production [3, 14,40]. Most of these soil conditions have led to desertification in some parts due to an increase in the soil erosion and decreases in the soil fertility due to the depletion of nutrients related to the loss of organic matter and the biological diversification of soils [41]. Bowker et al. [41] noted that most of the problems related to desertification in arid and semiarid environments originate from anthropogenic disturbances that change the organic matter and biological properties of soil crusts. This is especially true for arid regions, where desertification is closely related to either a net loss in the soil fertility or a redistribution of the soil nutrients [42–44]. Bowker et al. [41] found that the micronutrient soil content is highly correlated to the biological activity and specifically to the presence of moss and lichen species in semiarid conditions. The distribution of these biological species is highly correlated to the moisture, organic matter content, and higher availability of micronutrients. The high degradation observed in arid lands can therefore be resolved by improving the biological diversity of soil through the management of the organic matter.

From this point of view, biological diversity and the management of the organic matter become important factors for the improvement of micronutrient availability in degraded lands. In most cases, the response of plant growth to the addition of macronutrients and micronutrients in semiarid and arid soils is highly positive [43]. In fact, the response to the addition of a heterogeneous nutrient supply must be optimised in each case by understanding the plant response and the bioavailability of each nutrient [43]. To monitor the addition of different concentrations of nutrients and their effect on plant growth, we need to understand how the nutrient concentrations in the plant respond to changes in nutrient supply. The effects of nutrients on plants can be easily determined by the ability of plants to translocate nutrients and to store pools in different parts of the plant. Thus, the ideal management of the micronutrients will require knowledge on how the different nutrients are specifically taken up by different plants and how the nutrients are stored and translocated inside the different plant parts, particularly the edible portions [27].

Another important issue associated with the availability of different nutrients in soil is the speciation, quantity, and distribution of the nutrient in different soils, as well as the effects of different chemical species on root growth and grain yield [27]. Modern agricultural techniques used to obtain higher yields worldwide are accelerating the depletion of nutrients and leading to an increase in the application of fertilisers to maintain production [45]. Li et al. [46] found that the micronutrient status in soils and crops is affected by fertilisations practices. Specific cases must be studied to determine how fertilisation practices can help improve the soil micronutrient and macronutrient states in order to manage the ideal rates of fertiliser application for in different crops. Research on cereal crops showed that it is possible to obtain higher yields with the application of organic fertilisers because these improve the soil organic matter and thus provide available micronutrients for crops. However, low yields of winter wheat were obtained when only the micronutrient contents were improved because these were highly concentrated in the plant tissues and grains [46]. Sustainable crop production needs to maintain the soil fertility on a long-term basis, and it is essential that the organic matter and nutrients removed during harvest are replenished through external application on a regular basis. Some authors recommend the maintenance of the soil organic matter at a threshold level, which will depend on the soil and climatic factors, to ensure the physical, chemical, and biological integrity of soil to achieve sustainable agricultural and environmental functions [47–51].

In semiarid and arid environments, some studies concluded that the presence of Se in plants has a protective effect that helps maintain the water status in plants exposed to drought [25]. In this case, selenium is hypothesised to participate in the regulation of the water status of plants and in maintaining the content of water in the tissues at a sufficiently high level [25]. The selenium-supplemented fertilisation of wheat crops in semiarid conditions is a cost-effective method if inorganic fertilisers are used to ensure the ideal concentration of organic Se in grain as a result of dietary needs. Based on the research studies conducted on Se fertilisation as selenate (more mobile form of Se than selenite [52]) at different rates (4 to 200 g/ha), wheat does not express any phytotoxicity (critical tissue level of 325 mg/kg) compared with other crops (tobacco, soybean, and rice), and the crop exhibited an excellent increase in the grain Se content for human consumption. The results noted the possibility of the use of sodium selenate for the biofortification of wheat in Australia [20, 53, 54].

The incorporation of organic by-products as soil conditioners and fertilisers has the advantage of recycling the macronutrients and micronutrients, enhancing the soil conditions, helping establish a sustainable vegetative cover, avoiding erosion, maximising crop yield when applied with best management practices, and monitoring potential hazards to soils and crops [30, 55].

## 5. USE OF ORGANIC FERTILISERS TO IMPROVE SE CONTENT IN CROPS

Organic matter is able to increase the sorption of selenate in arid conditions and to act as a reservoir for crops to avoid the leaching, evaporation, and precipitation of this species to deeper soil layers [56]. Thus, it is recommended in the management of Se fertilisation to avoid the consumption of expensive and unsustainable inorganic forms. Organic matter plays a very important role in selenium immobilisation and the availability of selenium to plants. This effect depends on the compounds because organic matter includes highly heterogeneous compounds and the interactions are not very well understood. In this particular case, fulvic and humic acids appear to have an important role on the metalloid regulation through immobilisation and slow release of organic forms of Se [56].

According to Park et al. [57] biosolids and especially poultry and livestock manure are a good source of Se for crops. These organic amendments contain bioavailable forms of this nutrient and are a useful form of amendment since in most cases they are applied to the land as a form of reutilisation of an increasing "waste" product. The addition of manure and organic amendments to agricultural soils is the major source of most micronutrients including Se and sewage sludge has the same Se forms as manure [12].

According to Øgaard et al. [58], the addition of cattle manure to a loam soil in combination with selenite and selenate decreased the adsorption and immobilisation of both anions because the organic acids from the manure compete with selenium for the sorption sites. Organic fertilisers provide greater micronutrient content to plants not only because of the higher concentration of these but also due to their slower release from the plant [12, 25, 59].

In the case of recycling crop residues, such as straw, much of the added Se, that is, as much as 70–75%, remains in the rest of the plant, particularly in leaves, which are a highly concentrated source of this element. This is also true for animal manures because most of the Se is not absorbed by the animal and because farmed animals are fed feedstuffs rich in Se, when animals are fed with forage supplements or are raised on Se-rich soils. Thus, organic fertilisers provide a diverse, albeit unpredictable, source of Se.

Se in organic amendments has been proved to have several advantages compared with inorganic Se fertilizers. For example, Se is bioavailable through several microbial and chemical processes including immobilization, reduction, volatilization, and rhizosphere modification (Figures 4 and 5). The bioavailability depends on several factors, the most important of which is pH. In most of the case this availability is assured by the treatment of organic amendments, where bioavailability is increased by transforming Se from solid to the soil solution phase, improving the mobilisation of the metalloid [41] and increasing the pH buffering capacity.

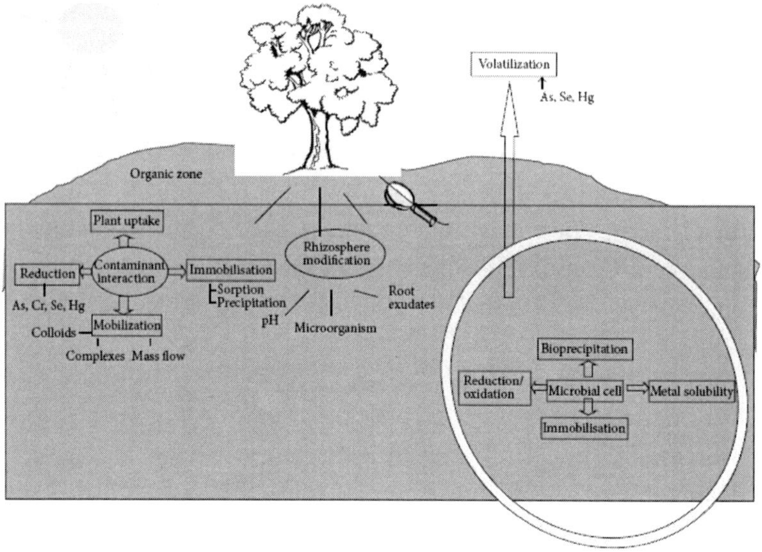


**Figure 4.** The role of organic amendments in regulating various bioremediation processes (extracted from [57]).

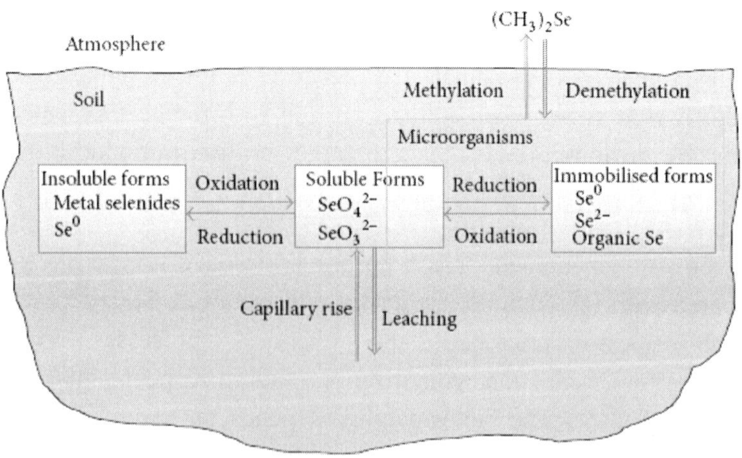


**Figure 5.** Cycle of Se in soil. Double arrows show preferred pattern for biodegradation (extracted from [62]).

Organic amendments to fertilise crop with Se offers an additional advantage in arid and semiarid environments since the removal of Se in topsoil is mostly due to volatilization [60, 61]. Indeed, according to Flury et al. [62], the elimination of Se from topsoil mostly happens due to volatilization and leaching, Figure 5. These authors investigated Se removal from organic amendments and summarised that when an abundance of organic C was available in the soil, microbial biodegradation of Se is activated through volatilisation. In general, leaching dominates in wet and cold conditions, whereas volatilization prevails in dry and warm conditions.

## 6. CONCLUSIONS

Recent studies on human health have shown that the appropriate amounts of certain micronutrients in the diet are important for maintaining a healthy society. The evidence for this is especially true in the case of selenium (Se), where 15% of the global population has an insufficiency. The deficiency of this element in the human and animal diets is related to some cancers and numerous physiological disorders. Many countries have implemented Se-supplemented fertilisation strategies, mainly for the production of cereals, because wheat is the most widely used food for improving the amounts of Se in the diet. The issues of Se availability is especially crucial in arid and semiarid environments, where one of the main benefits of the micronutrient to crop production is an increase in the drought resistance, and where many of the soils that are low in selenium are located.

There are constraints that must be taken into account regarding the application of Se-supplemented fertilisers. Mineral biofortification has been proven to be effective; however, this strategy is limited by the availability of mineral resources and its cost [34]. Sustainable biofortification depends on the use of organic fertilisers and conditioners to improve micronutrient fertilisation, particularly Se fertilisation. Thus, as well as providing other better-known soil benefits, the addition of organic matter could become an important consideration to ensure a healthy Se content in wheat produced in semiarid conditions.

We have reviewed research conducted in this field and found that further studies must be performed to compare the performance of different sources of organic matter as a fertiliser including biosolids and especially poultry and livestock manure, straw, or any organic by-product, to ensure the appropriate levels of available Se, as well as other micronutrients, in the fertilisation of wheat to improve the micronutrition of the human diet. More research must be conducted to understand the speciation and release of Se from organic amendments, to improve the use of organic by-products, and to avoid the limitations related to the use of inorganic fertilisers. Finally, research must be focused on the need of each specific land location and crop to provide optimised Se contents for animal and plant intake, avoiding the immobilisation of this metal in soils. Therefore, specific recommendations must be done for specific organic amendments and agricultural uses to optimise uptake and avoid toxicity.

## REFERENCES

1. P. J. White and M. R. Broadley, "Biofortification of crops with seven mineral elements often lacking in human diets—iron, zinc, copper, calcium, magnesium, selenium and iodine," New Phytologist, vol. 182, no. 1, pp. 49–84, 2009.

2. J. L. Stroud, M. R. Broadley, I. Foot et al., "Soil factors affecting selenium concentration in wheat grain and the fate and speciation of Se fertilisers applied to soil," Plant and Soil, vol. 332, no. 1, pp. 19–30, 2010.

3. D. Tennant, G. Scholz, J. Dixon, and B. Purdie, "Physical and chemical characteristics of duplex soils and their distribution in the south-west of Western Australia," Australian Journal of Experimental Agriculture, vol. 32, no. 7, pp. 827–843, 1992.

4. U. C. Gupta, W. Kening, and L. Siyuan, "Micronutrients in soils, crops and livestock," Earth Science Frontiers, vol. 15, no. 5, pp. 110–125, 2008.

5. G. Gissel-Nielsen, "Effects of selenium supplementation of field crops," in Environmental Chemistry of Selenium, W. T. Frankenberger and R. A. Engberg, Eds., pp. 99–112, Dekker, New York, NY, USA, 1998.

6. P. Ekholm, H. Reinivuo, P. Mattila et al., "Changes in the mineral and trace element contents of cereals, fruits and vegetables in Finland," Journal of Food Composition and Analysis, vol. 20, no. 6, pp. 487–495, 2007.

7. S. Mythili, K. Natarajan, and R. Kalpana, "Zinc nutrition in rice: a review," Agricultural Reviews, vol. 24, no. 2, pp. 136–141, 2003.

8. G. Bañuelos and Z. Q. Lin, Development and Uses of Biofortified Agricultural Products, CRC Press, Boca Raton, Fla, USA, 2009.

9. B. H. Robinson, G. Bañuelos, H. M. Conesa, M. W. H. Evangelou, and R. Schulin, "The phytomanagement of trace elements in soil," Critical Reviews in Plant Sciences, vol. 28, no. 4, pp. 240–266, 2009.

10. S. Hueso, T. Hernández, and C. García, "Resistance and resilience of the soil microbial biomass to severe drought in semiarid soils: the importance of organic amendments," Applied Soil Ecology, vol. 50, no. 1, pp. 27–36, 2011.

11. P. M. Haygarth, "Global importance and global cycling of selenium," in Selenium in the Environment, W. T. Frankenberger and B. Sally, Eds., pp. 1–28, Marcel Dekker, New York, NY, USA, 1994.

12. M. P. Rayman, "The argument for increasing selenium intake," Proceedings of the Nutrition Society, vol. 61, no. 2, pp. 203–215, 2002.

13. M. F. Fenech, "Dietary reference values of individual micronutrients and nutriomes for genome damage prevention: current status and a road map to the future," American Journal of Clinical Nutrition, vol. 91, no. 5, pp. 1438S–1454S, 2010.

14. G. W. Ford, J. J. Martin, P. Rengasamy, S. C. Boucher, and A. Ellington, "Soil sodicity in Victoria,"Australian Journal of Soil Research, vol. 31, no. 6, pp. 869–909, 1993.

15. M. R. Broadley, P. J. White, R. J. Bryson et al., "Biofortification of UK food crops with selenium,"Proceedings of the Nutrition Society, vol. 65, no. 2, pp. 169–181, 2006.

16. F. Fordice, "Selenium deficiency and toxicity in the environment," in Essentials of Medical Geology, O. Selinus, B. Alloway, J. Centeno et al., Eds., pp. 373–415, Elsevier, London, UK, 2005.

17. D. V. Frost, "What do losses in selenium and arsenic bioavailability signify for health?" Science of the Total Environment, vol. 28, pp. 455–466, 1983.

18. National Academy of Sciences, Recommended Dietary Allowances, National Academy of Sciences, Washington, DC, USA, 10th edition, 1989.

19. J. Lee, D. G. Masters, C. L. White, N. D. Grace, and G. J. Judson, "Current issues in trace element nutrition of grazing livestock in Australia and New Zealand," Australian Journal of Agricultural Research, vol. 50, no. 8, pp. 1341–1364, 1999.

20. A. D. Lemly, "Guidelines for evaluating selenium data from aquatic monitoring and assessment studies,"Environmental Monitoring and Assessment, vol. 28, no. 1, pp. 83–100, 1993.

21. O. A. Levander and R. F. Burk, "Uptake of human dietary standards for selenium," in Selenium Its Molecular Biology and Role in Human Health, D. L. Hatfield, M. J. Berry, and V. N. Gladyshev, Eds., pp. 399–410, Springer, New York, NY, USA, 2nd edition, 2006.

22. G. Lyons, "Selenium in cereals: improving the efficiency of agronomic biofortification in the UK," Plant and Soil, vol. 332, no. 1, pp. 1–4, 2010.

23. J. Wu, C. Salisbury, R. Graham, G. Lyons, and M. Fenech, "Increased consumption of wheat biofortified with selenium does not modify biomarkers of cancer risk, oxidative stress, or immune function in healthy Australian males," Environmental and Molecular Mutagenesis, vol. 50, no. 6, pp. 489–501, 2009.

24. K. Schwarz and C. M. Foltz, "Selenium as an integral part of factor 3 against dietary necrotic liver degeneration," Journal of the American Chemical Society, vol. 79, no. 12, pp. 3292–3293, 1957.

25. K. Kaur, R. K. Jalota, and D. J. Midmore, "Impact of tree clearing on soil attributes for a pastoral property in central Queensland, Australia," Soil Science, vol. 172, no. 7, pp. 516–533, 2007.

26. K. M. Havstad, J. E. Herrick, and W. H. Schlesinger, "Desert rangelends, degradation and nutrients," inRangeland Desertification, O. Arnalds and S. Archer, Eds., pp. 77–87, Kluwer Academic Publishers, Dordrecht, The Netherlands, 2000.

27. E. A. Pilon-Smits, C. F. Quinn, W. Tapken, M. Malagoli, and M. Schiavon, "Physiological functions of beneficial elements," Current Opinion in Plant Biology, vol. 12, no. 3, pp. 267–274, 2009.

28. C. D. Thomson, "Assessment of requirements for selenium and adequacy of selenium status: a review,"European Journal of Clinical Nutrition, vol. 58, no. 3, pp. 391–402, 2004.

29. M. P. Rayman, "Food-chain selenium and human health: emphasis on intake," British Journal of Nutrition, vol. 100, no. 2, pp. 254–268, 2008.

30. G. J. Judson and D. J. Reuter, Soil Analysis: An Interpretation Manual, South Australia Research & Development Institute (SARDI), Urrbrae, South Australia, 1998.

31. G. H. Lyons, J. C. R. Stangoulis, and R. D. Graham, "Tolerance of wheat (Triticum aestivum L.) to high soil and solution selenium levels," Plant and Soil, vol. 270, no. 1, pp. 179–188, 2005.

32. J. D. Rosen, "A Review of the nutrition claims made by proponents of organic food," Comprehensive Reviews in Food Science and Food Safety, vol. 9, no. 3, pp. 270–277, 2010.

33. K. Nakamuro, K. Nakanishi, T. Okuno, T. Hasegawa, and Y. Sayato, "Comparison of methylated selenium metabolites in rats after oral administration of various selenium compounds," Japanese Journal of Toxicology and Environmental Health, vol. 43, no. 1, pp. 1482–1489, 1997.

34. T. Hasegawa, M. Mihara, K. Nakamuro, and Y. Sayato, "Mechanisms of selenium methylation and toxicity in mice treated with selenocystine," Archives of Toxicology, vol. 71, no. 1-2, pp. 31–38, 1996.

35. Y. G. Zhu, E. A. H. Pilon-Smits, F. J. Zhao, P. N. Williams, and A. A. Meharg, "Selenium in higher plants: understanding mechanisms for biofortification and phytoremediation," Trends in Plant Science, vol. 14, no. 8, pp. 436–442, 2009.

36. G. Lyons, I. Ortiz-Monasterio, J. Stangoulis, and R. Graham, "Selenium concentration in wheat grain: is there sufficient genotypic variation to use in breeding?" Plant and Soil, vol. 269, no. 1-2, pp. 369–380, 2005.

37. M. H. Eurola, P. I. Ekholm, M. E. Ylinen, P. E. Koivistoinen, and P. T. Varo, "Selenium in finish foods after beginning the use of selenate-supplemented fertilisers," Journal of the Science of Food and Agriculture, vol. 56, pp. 57–70, 1991.

38. M. R. Broadley, J. Alcock, J. Alford et al., "Selenium biofortification of high-yielding winter wheat (Triticum aestivum L.) by liquid or granular Se fertilisation," Plant and Soil, vol. 332, no. 1, pp. 5–18, 2010.

39. M. Eurola, V. Hietaniemi, M. Kontturi et al., "Selenium content of Finnish oats in 1997–1999: effect of cultivars and cultivation techniques," Agricultural and Food Science, vol. 13, no. 1-2, pp. 46–53, 2004.

40. I. Bertrand, R. E. Holloway, R. D. Armstrong, and M. J. McLaughlin, "Chemical characteristics of phosphorus in alkaline soils from southern Australia," Australian Journal of Soil Research, vol. 41, no. 1, pp. 61–76, 2003.

41. M. A. Bowker, J. Belnap, D. W. Davidson, and S. L. Phillips, "Evidence for micronutrient limitation of biological soil crusts: importance to arid-lands restoration," Ecological Applications, vol. 15, no. 6, pp. 1941–1951, 2005.

42. H. Hartikainen, "Biogeochemistry of selenium and its impact on food chain quality and human health,"Journal of Trace Elements in Medicine and Biology, vol. 18, no. 4, pp. 309–318, 2005.

43. A. D. Robson, N. E. Longnecker, and L. D. Osborne, "Effects of heterogeneous nutrient supply on root growth and nutrient uptake in relation to nutrient supply on duplex soils," Australian Journal of Experimental Agriculture, vol. 32, no. 7, pp. 879–886.

44. W. H. Schlesinger, J. F. Reynolds, G. L. Cunningham et al., "Biological feedbacks in global desertification," Science, vol. 247, no. 4946, pp. 1043–1048, 1990.

45. Z. Rengel, G. D. Batten, and D. E. Crowley, "Agronomic approaches for improving the micronutrient density in edible portions of field crops," Field Crops Research, vol. 60, no. 1-2, pp. 27–40, 1999.

46. H. F. Li, S. P. McGrath, and F. J. Zhao, "Selenium uptake, translocation and speciation in wheat supplied with selenate or selenite," New Phytologist, vol. 178, no. 1, pp. 92–102, 2008.

47. P. Pathak, K. L. Sahrawat, S. P. Wani, R. C. Sachan, and R. Sudi, "Opportunities for water harvesting and supplemental irrigation for improving rainfed agriculture in semi-arid areas," in Rainfed Agriculture: Unlocking the Potential. Comprehensive Assessment of Water in Agriculture Series, S. P. Wani, J. Rockström, and T. Oweis, Eds., vol. 7, pp. 197–221, CAB International, Wallingford, UK, 2009.

48. A. Bationo, J. Kihara, V. Vanlauwe, J. Kimetu, B. S. Waswa, and K. L. Sahrawat, "Integrated nutrient managemenr: concepts and experience form Sub-Saharan Africa," in Integrated Nutrient Management for Sustainable Crop Production, M. S. Auklakh and C. A. Grant, Eds., pp. 467–521, The Haworth Press-Taylor and Francis, New York, NY, USA, 2008.

49. A. D. Sparrow, M. H. Friedel, and D. J. Tongway, "Degradation and recovery processes in arid grazing lands of central Australia. Part 3: implications at landscape scale," Journal of Arid Environments, vol. 55, no. 2, pp. 349–360, 2003.

50. N. E. Spencer and S. M. Siegel, "Effects of sulfur and selenium oxyanions on Hg-toxicity in turnip seed germination," Water, Air, and Soil Pollution, vol. 9, no. 4, pp. 423–427, 1978.

51. S. H. Van Dorst and P. J. Peterson, "Selenium speciation in the soil solution and its relevance to plant uptake," Journal of the Science of Food and Agriculture, vol. 35, pp. 601–605, 1984.

52. V. V. Kuznetsov, V. P. Kholodova, V. V. Kuznetsov, and B. A. Yagodin, "Selenium regulates the water status of plants exposed to drought," Dokaldy Biological Sciences, vol. 390, no. 1–6, pp. 266–268, 2003.

53. M. A. Elrashidi, D. C. Adriano, and W. L. Lindsay, "Solubility, speciation and transformations of selenium in soils," in Selenium in Agriculture and the Environment, L. W. Jacobs, Ed., Special Publication Number 23, pp. 51–63, SSSA, Madison, Wis, USA, 1989.

54. M. S. Fan, F. J. Zhao, P. R. Poulton, and S. P. McGrath, "Historical changes in the concentrations of selenium in soil and wheat grain from the Broadbalk experiment over the last 160 years," Science of the Total Environment, vol. 389, no. 2-3, pp. 532–538, 2008.

55. R. K. Bastian, "Interpreting science in the real world for sustainable land application," Journal of Environmental Quality, vol. 34, no. 1, pp. 174–183, 2005.

56. A. Fernández-Martínez and L. Charlet, "Selenium environmental cycling and bioavailability: a structural chemist point of view," Reviews in Environmental Science and Biotechnology, vol. 8, no. 1, pp. 81–110, 2009.

57. J. H. Park, D. Lamb, P. Paneerselvam, G. Choppala, N. Bolan, and J. W. Chung, "Role of organic amendments on enhanced bioremediation of heavy metal(loid) contaminated soils," Journal of Hazardous Materials, vol. 185, no. 2-3, pp. 549–574, 2011.

58. A. F. Øgaard, T. A. Sogn, and S. Eich-Greatorex, "Effect of cattle manure on selenate and selenite retention in soil," Nutrient Cycling in Agroecosystems, vol. 76, no. 1, pp. 39–48, 2006.

59. M. P. Rayman, H. G. Infante, and M. Sargent, "Food-chain selenium and human health: spotlight on speciation," British Journal of Nutrition, vol. 100, no. 2, pp. 238–253, 2008.

60. W. T. Frankenberger Jr. and U. Karlson, "Soil management factors affecting volatilization of selenium from dewatered sediments," Geomicrobiology Journal, vol. 12, no. 4, pp. 265–278, 1994.

61. W. T. Frankenberger Jr. and U. Karlson, "Volatilization of selenium from a dewatered seleniferous sediment: a field study," Journal of Industrial Microbiology, vol. 14, no. 3-4, pp. 226–232, 1995.

62. M. Flury, W. T. Frankenberger Jr., and W. A. Jury, "Long-term depletion of selenium from Kesterson dewatered sediments," Science of the Total Environment, vol. 198, no. 3, pp. 259–270, 1997.

# CHAPTER 12

# Climate variability and water resource scarcity in drylands of Rajasthan, India

**Ram Babu Singh, Ajay Kumar**

[1] Department of Geography, Delhi School of Economics, University of Delhi, Delhi 110007, India.

[2] Department of Geography, SSN College, University of Delhi, Delhi 110036, India.

## ABSTRACT

### Background

The world is experiencing a high climatic variability and has its impact on all ecosystems, regions and sectors, but the nature and extent of vulnerability to climate change differs with adaptive capacity of individual, society and region. The regions which are scarce on resources are among the very vulnerable systems of the earth and most of such regions have existing extreme climatic conditions like deserts and snow covered areas. The vulnerability of arid regions is further accentuated by low levels of socio-economic development, and this is the case with arid regions of India, which leads to fast depletion of water resources.

## Results

The study analyses the climate variability over last 35 years (1973–2008) in the Western drylands of Rajasthan and the results show that the temperature has shown a great variability with an average rise of about 0.5°C for the month of June. The maximum rise is visible during 1995–2000. The scarcity of water resource accompanied by variability creates water stress conditions in the region and is responsible for droughts. During last 35 years (1973–2008) the average annual rainfall has shown a decrease of 50 mm. With an overall rise in temperature over the region and changing precipitation trends the humidity provinces as defined by Thornthwaite have shifted eastward.

## Conclusions

The climate variability has direct impact on the water resources of the region. The low to medium rainfall in most parts, high evapo-transpiration demands for water, high frequency of occurrence of droughts resulting from the departure of rainfall from mean values, and the high per capita arable land increases the demand for irrigation water. The free power in agriculture continued for many years, and the existing pump horse power based pricing of electricity encourage over-pumping and inefficient and often wasteful use of groundwater. The advent of Indira Gandhi Canal in the fragile desert ecosystems helps the region to cope with high variability in rainfall. The study suggests that water security in the region can be attained through expansion of canal network.

## KEYWORDS

Climate variability; Arid regions; Water scarcity; Livelihood security

# 1. BACKGROUND

The world is experiencing a high variability in climatic parameters. The variability is both spatial and temporal in nature. A 10 to 15 per cent increase in monsoon precipitation in many regions of India, a simultaneous precipitation decline of 5–25 per cent in drought-prone central India and a sharp decline in winter rainfall in northern India are projected (Ramesh and Yadava, 2005). Possible changes in extreme weather events may cause increase in frequency of heat waves and drought in the northern Indian region (Sen Roy and Singh, 2002). The impact on water balance, hydrology and vegetation are also predicted to be severe. Climate variability has its impact on all ecosystems, regions and sectors, but the nature and extent of vulnerability to climate change differs with adaptive capacity of individual, society and region. The regions which are scarce on resources are among the very vulnerable systems of the earth and most of such regions have existing extreme climatic conditions like deserts and snow covered areas (Watson et al., 1996; Ribot Jesse 1996). Droughts, desertification, and water shortages are permanent features of life in drylands and under such conditions the increase in evapotranspiration has severe impact on agriculture, horticulture and forestry and human activities (Sharma and Bhattacharya 2006; Smith et al., 1996; Sen Roy and Singh, 2002). In an area dominated by arid and semi-arid lands, water is a very limited and high percentage of degraded land is there due to Aeolian processes and because soils tend to be thin, much of the rainfall runs directly off of the surface, only to infiltrate deeper soils downslope or along river beds, changes in the frequency and magnitude of rainfall events will alter the number of recharge events (Intergovernmental Panel on Climate Change 1996). The vulnerability of arid regions is further accentuated by low levels of socio-economic development, and this is the case with arid regions of India (Singh and Gurjar, 2011). The high population density of Indian desert is intensifying the pressure on the land and water resources, which leads to increased rate of land degradation and fast depletion of water resources (Kharin, 1995). Beside population pressure, climate variability also has very important role in determining the status of resources. Climate change is expected to increase the frequency and intensity of current hazards and the probability of extreme events, and also to spur the emergence of new hazards (Nicholls and Lowe, 2006).

Extreme events within the variability of the climate system are, by far, the largest cause of natural disasters worldwide each year (Kininmonth, 2004).

**Table 1.** Account of withdrawal and recharge of groundwater in western Rajathan

| Year | 1984 | 1990 | 1995 | 1998 | 2001 | 2004 |
|---|---|---|---|---|---|---|
| A. Gross groundwater recharge | 16.22 | 12.71 | 13.16 | 12.6 | 11.16 | 11.56 |
| 1. Irrigation | 4.93 | 5.42 | 9.09 | 11.04 | 10.45 | 11.6 |
| 2. Domestic and Industrial | 2.11 | 1.99 | 0.7 | 0.98 | 1.18 | 1.39 |
| B. Gross Draft (1 + 2) | 7.04 | 7.42 | 9.78 | 12.02 | 11.64 | 12.99 |
| Groundwater balance (A-B) | 9.19 | 5.29 | 3.38 | 0.58 | −0.48 | −1.43 |
| Stage of groundwater development (%) | 36 | 54 | 59 | 69 | 104 | 125 |

Source: Groundwater Board, Rajasthan, 2009.

**Table 2.** Status of groundwater in the western Rajathan

| Year | Block (area unit) | Safe (white) | Semi-critical (semi-gray) | Critical (gray) | Over-exploited (dark) |
|---|---|---|---|---|---|
| 1984 | 236 | 203 (86) | 10(4.2) | 11 (4.7) | 12 (5.1) |
| 1988 | 236 | 122 (54) | 42(18.6) | 18 (8) | 44 (19.4) |
| 1990 | 236 | 148 (62.7) | 31 (13.1) | 13 (5.5) | 44 (19.4) |
| 1992 | 236 | 149 (63.1) | 19 (8.1) | 15 (6.4) | 53 (22.5) |
| 1995 | 236 | 127 (53.8) | 35 (14.8) | 14 (5.9) | 60 (25.4) |
| 1998 | 236 | 135 (57.9) | 34 (14.6) | 23 (9.9) | 41 (17.6) |
| 2001 | 236 | 49 (20.8) | 21 (8.9) | 80 (33.9) | 86 (36.4) |
| 2004 | 237 | 32 (14) | 14 (6) | 50 (21) | 140 (59) |
| 2009 | 236 | 30 (13) | 8 (3) | 34 (14) | 164 (69) |

Source: Groundwater Board, Rajasthan, 2009.

Whatever might be the nature of climate change that bounds to occur in a region, the outcome would be large scale displacement of normal human activity. And, now the climate is changing due to human actions. Despite efforts to abate the human causes, human-driven climate change will continue for decades and longer (Intergovernmental Panel on Climate Change IPCC 2001). Global climate change, which has already impacted natural resource systems across the Earth, is adding to the pressures and is expected to substantially disrupt many of these systems and the goods and services that they provide (Millennium Ecosystem Assessment, 2005). Therefore, only climate change mitigation steps are not enough; there is an immediate need to strengthen the adaptive capacity of people and regions. Therefore in lieu of these issues the present study attempts to analyse the status and vulnerability of water resources in the western drylands of Rajasthan in context of climate variability over past 35 years (1973–2008).

## 2. STUDY AREA

The state of Rajasthan is the second largest state in India, with an area of 0.342 million sq km out of which 0.198 million sq km which is nearly 58 per cent, is occupied by Thar Desert (Figure 1). The Thar Desert is characterized by low and erratic rainfall, high evapotranspiration, large temperature variation, scanty vegetation, rodent infestation, absence of perennial rivers, sparse and nomadic population and dependence of man on animal rearing. The study area is located in the north-western part of India between 24°31′ to 30°12′ north latitudes and 69°15′ to 76°42′ east longitudes. It is surrounded by Punjab in north, Gujarat in south, Pakistan in west and Aravalis in east. Its length is 640 km from north to south and width is 300 km from east to west. The western arid region includes 12 districts i.e. Sriganganagar, Hanumangarh, Churu, Sikar, Jhunjhunu, Nagaur, Bikaner, Jodhpur, Barmer, Jalore, Pali and Jaisalmer. The study area can be divided into two main physical regions viz. i) Sandy Arid Region and ii) Semi-Arid. The western sandy plain includes the *Marusthali* and the adjoining *Bangar* to the west of the Aravalis. It is a wide expanse of poorly watered, sterile and wind-blown sand. The general aspect of this region is of an interminable sea of sand and sand dunes of different shapes and sizes varying from 6–60 meter in height and being sometimes 3–5 km in length (Figure 2). Semi-Arid region is drained by Luni in its south-eastern portion. Gullying has given rise to conglomerate landscape. The land is slightly undulating within venue of sand deposited by inland drainage and stream with salt lakes (Sharma and Sharma, 2002) (Figure 3).The Western arid region of Rajasthan has a population of 22.5 million as per census data of 2001. The Indian Thar Desert is most populous desert in the world, with density of around 84 persons per sq km. The population growth (42 per cent) is also very high between 1901 and 2001, and more than 70 per cent people lives in villages or scattered settlement called *dhanis* (Hamlets) where agriculture and animal husbandry are the main occupation.

# 3. RESEARCH DATA AND METHODOLOGY

The secondary data sources include historical climate data of 35 years from 1973 to 2008 of WMO standard in respect of rainfall, maximum temperature, minimum temperature and rainy days from India Meteorological Department (IMD). The temporal and spatial analysis of data has been done by using Moving average (Seven years). This tool is used to find out the trend of temperature and rainfall variation over the years. Thornthwaite and Mather Moisture index (1955) is useful in division of climate of a year into various types. The comparison of moisture indices of each month of this year with previous years is helpful in determining changes in climate for particular months in a year. Maps are used to determine the spatio-temporal change in temperature and precipitation. Arc GIS 10.0, Erdas 9.2 and Arc View 3.2 software were used for this purpose.

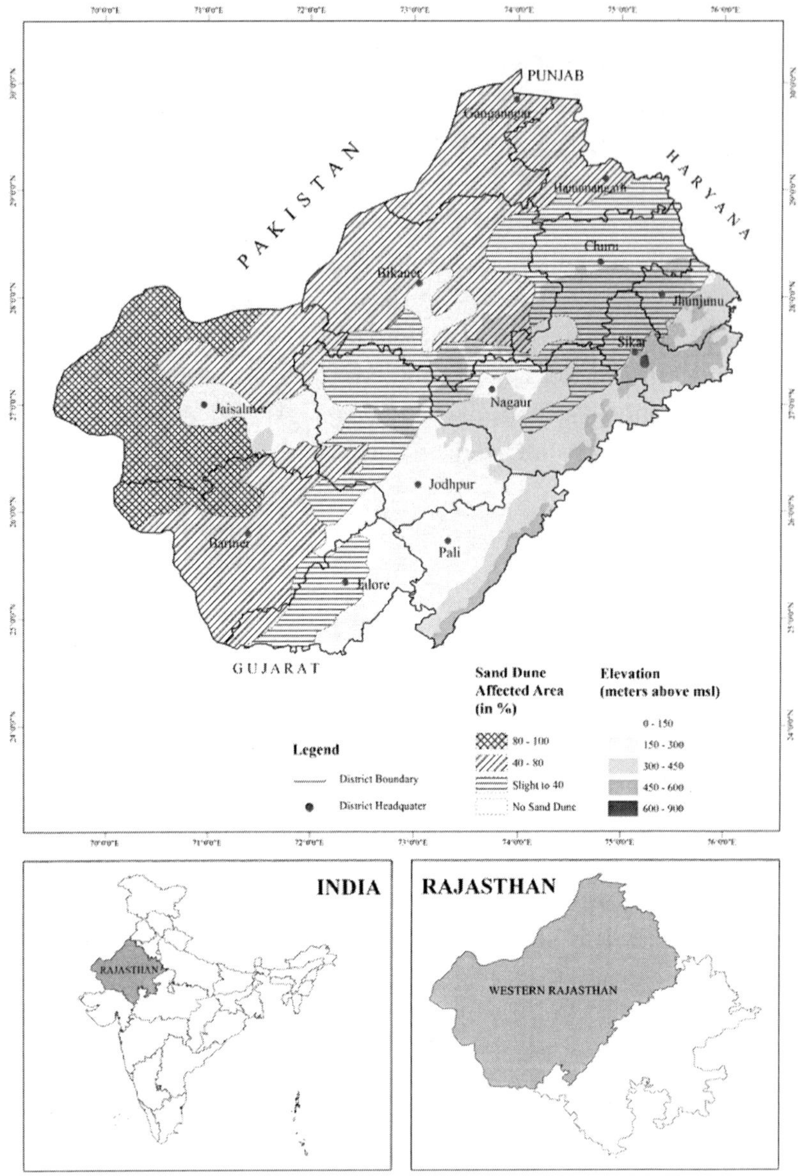


**Figure 1.** Location, relief and sand dune affected area of the study region.

**Figure 2.** Sand dunes in western sandy region.

**Figure 3.** Undulating topography of eastern semi-arid region.

# 4. THORNTHWAITE'S MOISTURE INDEX AND CLIMATE CLASSIFICATION

In 1948, Thornthwaite proposed a new classification of climate. The classification is based on the concept of potential evapotranspiration, which represents the amount of moisture that would be transferred to the atmosphere by evaporation of liquid or solid water plus transpiration from living tissues, principally plants if it (the moisture) were available. The potential evapotranspiration (PE) is calculated from the mean monthly temperature (in °C), with corrections for day length. For a 30-day month (12-hour days):

$$PE\ (\text{in cm}) = 1.6\ (10\ t/I)^{a}$$

where I = the sum of 12 months of $(t/5)^{1.514}$

$$a = (6.75e^{-7} \times I^{3}) - (7.71e^{-5I2} + 1.79e^{-2I} + 0.49239).$$

The monthly water surplus (S) or deficit (D) is calculated from a moisture budget assessment including stored soil moisture. Thornthwaite (1948) defined the humidity province based on moisture index. A moisture index (Im) is given by the following formula:

$$IM = (100S - 60D)PE$$

Land use map is based on AWiFS sensor data taken from National Remote Sensing Agency. Secondary data pertaining to land use and cropping pattern has been collected from Directorate of Agriculture, Rajasthan and using the Weaver's method, crop diversification and combination regions have been identified. The two time data related to crop area and production has been compared and results have been analysed. Data related to water resources have been taken from Water Resource Department, Government of Rajasthan and have been analysed using different statistical methods.

# 5. RESULTS AND DISCUSSION

## 5.1 Climate change

### 5.1.1 Temperature variability

In the Western Rajasthan during last 30 years the temperature has shown a great variability with an average rise of about 0.5°C for the month of June. The temperature rose from average temperature of 33.25°C in 1973 to 33.75°C in 2008. The maximum rise is visible during 1995–2000. Whereas for the month of January the average rise in temperature is about 1°C during last 30 years. The average monthly temperature for January has increased from 15°C in 1973 to 16°C in 2008 (Figure 4). The temperature rise for January is more consistent and the fluctuations are not so abrupt as compared to June. The local factor influencing the temperature rise is increased cloud cover during winter months.

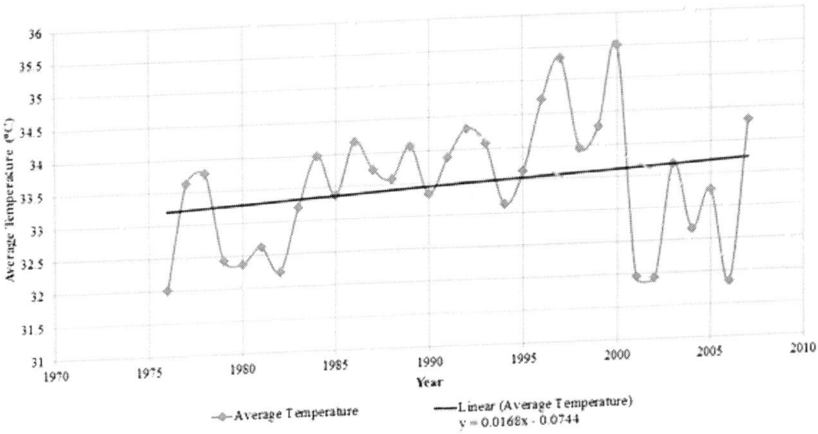


**Figure 4.** Temperature variation over past 35 years in Western Rajasthan during the month of January.

### 5.1.2 Rainfall variability

The Western Rajasthan suffers from scarcity of rainfall. The average annual rainfall in the region is measured to be 350 mm per year and above that the rainfall pattern in the region shows large variability and has never been consistent (Figure 5). The scarcity accompanied by variability creates water stress conditions in the region and is responsible for droughts. During last 30 years the average annual rainfall has shown a decrease of 50 mm, it was measured 400 mm during 1973, which came down to 350 mm per annum. The decrease in precipitation can be attributed to rising temperature. As the temperature increases the moisture holding capacity of atmosphere increases and thus the condensation is delayed till further addition of moisture to the atmosphere which might lead to decrease in precipitation.

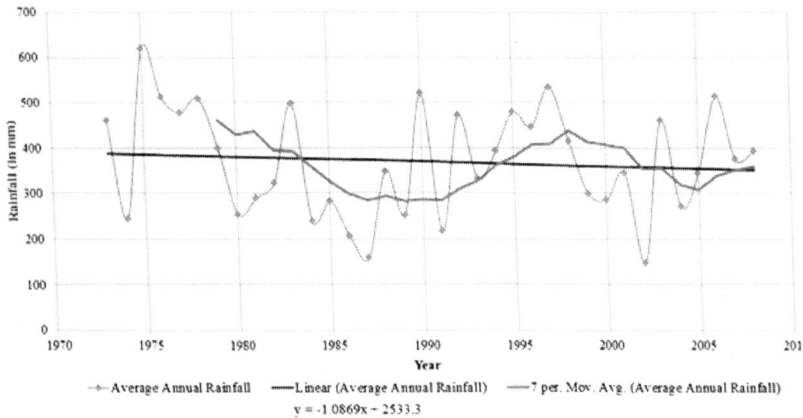


**Figure 5.** Variability in rainfall over past 35 years in Western Rajasthan.

## 5.2 Change in humidity provinces

With an overall rise in temperature over the region and changing precipitation trends the humidity provinces as defined by Thornthwaite have shifted eastward. The results from correlation and regression matrix show that the changes in PE, which further led to change in humidity provinces, are caused by change in maximum temperature. The influence of change in minimum temperature and precipitation on PE is less as compared to impact of change in maximum temperature. In 1973 the arid region consisted of Barmer, Bikaner, Ganganagar along with western parts of Churu, Hanumangarh, Jalore and Jodhpur, But according to 2008 meteorological datasets eastern parts of Churu, Hanumangarh and Jodhpur, and western parts of Jhunjhunu, Nagaur, Pali and Sikar have been added to arid type of humidity province. The semi-arid type of province did not include eastern margins of Pali and Sikar, but now the earlier recorded dry sub humid type of conditions in these two parts of region are replaced by semi-arid conditions (Figure 6). The region has experienced changes in moisture index, but the changes are not homogenous all over the region. Increase in aridity i.e. fall in index values have been noticed over major part of the region. Whereas a small part including of western Bikaner, northern Jaisalmer, central Barmer and western Jalore have experienced decrease in arid conditions. The increase in aridity ranges from a minimum of zero point to maximum of 28 points.

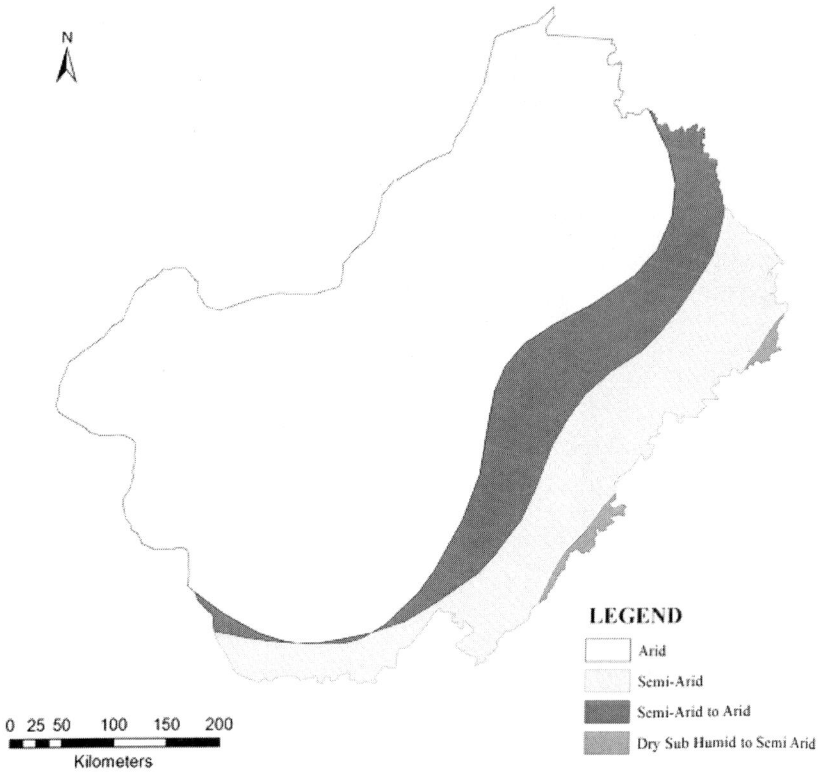


**Figure 6.** Shift of Thornthwaite's climate belts during 1970 – 2008.

## .3 Water resources

The region is characterized by heterogeneity in groundwater conditions. The region has all formations, viz., unconsolidated, semi consolidated and fully consolidated, with varying groundwater potential. Geographically, unconsolidated extensive porous formations, with low groundwater potential cover the northern, eastern and south western parts of the region, comprising the entire Ganganagar, most parts of Churu, Pali and Barmer districts. But, there are some pockets within these unconsolidated formations which have high yields (Figures 7 and 8). These patches are in Jhunjhunu and Sikar districts. But, the groundwater underlying the entire Ganganagar and Churu districts is saline, and therefore unfit for irrigation and drinking. Unconsolidated discontinuous aquifers with poor yield potential cover the western parts covering the Thar Desert in Jaisalmer. Most of it is saline, except some patches in the extreme west. Semi consolidated aquifers of low yield potential are found in parts of Jaisalmer, Bikaner and Churu districts. There are patches of semi consolidated aquifers with moderate yield potential in Barmer district, and high yield potential in the lower north western parts, covering parts of Bikaner and Churu districts. These aquifers are inherently saline. But, over the years, the quality of groundwater in this region has improved (Kumar et al., 2009).

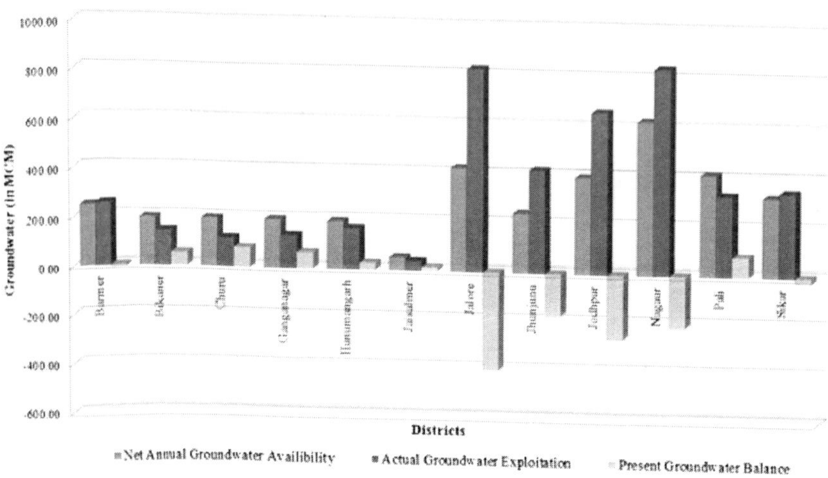


**Figure 7.** Groundwater resources potential of Western Rajasthan.

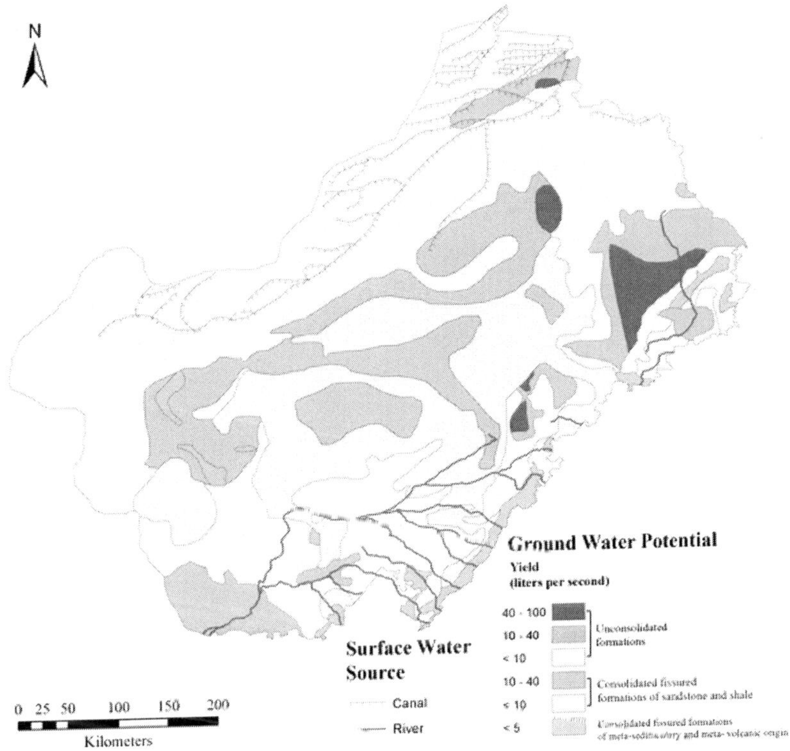


**Figure 8.** Surface water resources and groundwater potential in Western Rajasthan.

Consolidated fissured formations of sandstone and shale with low yield are found in Nagaur and Jodhpur districts, and that with moderate yield potential are found in other parts of Nagaur, Jodhpur and Jaisalmer districts. Consolidated limestone and dolomite formations in small patches are found in Nagaur, Jodhpur and Jaisalmer districts. Consolidated fissured formations of meta-sedimentary and meta-volcanic origin with low yield potential (1–5 lps) are found in the southern parts extending up to the central eastern part of region. They cover parts of Jalore district. Consolidated fissured sedimentary formations with low yield potential are found in parts of Jalore (Ozha and Sharma,2011).

Rajasthan stands first in terms of degree and extent of over-exploitation of groundwater resources in the country. One reason for this phenomenon is the absence of sufficient number of large-scale surface irrigation facilities, well-spread geographically. The low to medium rainfall in most parts, high evapo-transpiration demands for water, high frequency of occurrence of droughts resulting from the departure of rainfall from mean values, and the high per capita arable land increases the demand for irrigation water. This is being met through mining of groundwater resources. The free power in agriculture continued for many years, and the existing pump horse power based pricing of electricity encourage over-pumping and inefficient and often wasteful use of groundwater.

The irrigated area by all sources was only 9.3 per cent in 1951 has increased by about 43 per cent in 2000–01 out of the total potentiality (53 per cent) of irrigation in the western Rajasthan. This expansion of irrigation has resulted inclusion of new lands, which were earlier under fallow lands, pastures and grazing and land not available for cultivation and further it caused shrinkage of pastures and grazing lands. The data reveals that depletion in groundwater level is very significant in the region. Out of total 12 districts, 10 districts in the region shows depleting trend of groundwater level. On the basis of average depletion these districts have been further classified as most critical (>10 m), critical (average depletion is between 5 to 10 m) and moderate (average depletion is between 0 to 5 m) (Figure 9, Tables 1 and 2).

Overall, the estimated groundwater resources in Western Rajasthan are limited due to deep aquifers and low recharge. Based on the ratio of annual pumping to annual recharge, the stages of groundwater development have been categorized as safe (<70 per cent), semi-critical (70–90 per cent), critical (90–100 per cent) or overexploited (per cent 100 %) (Rathore, 2005; Narain et al.,2005). The groundwater is overexploited in Barmer, Jalore, Jhunjhunu, Jodhpur, Nagaur, and Sikar districts.

The advent of Indira Gandhi Canal in the northern parts of arid region of Rajasthan has led to immense change in the land use pattern of the region. The once barren land has been now converted into highly productive green belt (Figure 10). Such initiatives in the fragile desert ecosystems help the region to cope with high variability in rainfall. The scenario of land use had markedly changed during the years due to the canal irrigation. Satellite-based mapping has revealed that in arid western Rajasthan irrigated and double-cropped area has significantly increased and at the same time, the area under fallow land has declined. More area under groundwater irrigation with high inputs is coming under cash crops like groundnut, cotton, mustard, cumin and chillies. All these factors have added to the process of land degradation and has also increased the water demand in the region. It is important to note that area under cultivable wasteland, pastures and trees, fallow land and land not available for cultivation have declined by 24.07 per cent, 13.15 per cent and 8.55 per cent respectively in the western region. This trend of gradual decrease shows that the area that was earlier under open grazing for indigenous cows and other draught animals is shrinking rapidly, placing the traditional livelihood options under threat. With advent of canal irrigation the farmers prefer commercial and food crops rather than fodder crops. During 1960's there was dominance of pastures, grazing lands and areas of fodder crops. There is a decreasing trend of food grain crops in the region. Crop diversification has increased in present for both the cropping seasons as compared to 1960. One of the main driving forces behind crop diversification is availability of water through various sources viz. canal and tube wells. The area under fodder crops was 56 per cent in 1951, has reduced to 47 per cent in 1992–93 and about 36 per cent in 2003–04. The area under oil seeds cultivation increased from 6.2 per cent in 1951 to 21.4 per cent in 2003–04.

Besides all the physical factors, poor socio-economic conditions also accentuate the vulnerability. The region has highest population density (84 persons per sq km) among all arid regions of the world. The state of Rajasthan ranks low on development and literacy among the other states of India. The variability in population distribution also hampers the uniform planning, as according to the 2001 Census, Jaisalmer district, located on the western border, had a population of about half a million, accounting for about 2 per cent of total population of the region and more than 15 per cent of its area. On the other hand, Jhunjhunu district accounted for around 9 per cent of the state's population, but less than 1 per cent of its area. Therefore in order to attain sustainability and water security socio-economic conditions also need to be taken into consideration.

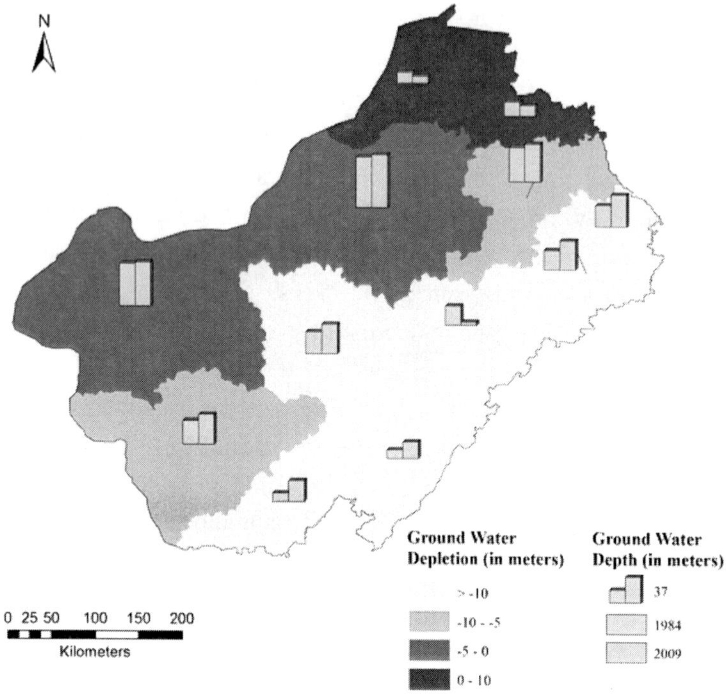


**Figure 9.** Groundwater depth and depletion in Western Rajasthan.

**Figure 10.** Agriculture amidst sand dunes due to canal irrigation.

# 6. CONCLUSION

The climate over Western Rajasthan is showing definite signs of change, with decrease in rainfall and increase in temperature and aridity. The arid climate belt has shifted eastward, intensifying the process of land degradation and causing desertification. The water resources are scarce, adding to vulnerability of region towards climate change. But in last few decades there has been tremendous development of various water sources by government in the region and due to these efforts the north-western part of region has attained good agriculture production and yield. The development of canal irrigation in some parts has intensified the process of land degradation in form of salinization and alkalization, which need to be managed at priority basis in order to maximize the profit from canal. The land use pattern in the region has also changed with development of water resources. Farmers are now opting for crop diversification and commercial crops like mustard. The decrease in rainfed crops like gram can be attributed to high variability of rainfall. The central part of the region is more dependent on rainfall for agriculture which makes it highly susceptible towards impact of climate change. There is a need to develop more extensive network of canals and water harvesting structures together with diversifying the source of livelihood.

# AUTHORS' CONTRIBUTIONS

RB S has been instrumental in conception and design of the study. He significantly contributed to interpretation of data. He has been involved in drafting the manuscript for important intellectual content. AK has acquired the data and has substantially contributed to analysis and interpretation of data. He has done the initial drafting of the manuscript. He prepared all the maps and graphs required for the study. Both author read and approved the final manuscript.

## REFERENCES

1. Intergovernmental Panel on Climate Change (1996) Climate Change 1995: IPCC Second Assessment Report. Intergovernmental Panel on Climate Change. Cambridge University Press, Cambridge

2. Intergovernmental Panel on Climate Change (IPCC) (2001) Climate Change 2001: Impacts, Adaptation Vulnerability. Contribution of Working Group II to the Third Assessment Report. Intergovernmental Panel on Climate Change, Geneva

3. Kharin NG (1995) Change in Biodiversity in Ecosystems of Central Asia under the Impacts of Desertification. United Nations, Washington DC

4. Kininmonth W (2004) Climate Change: A Natural Hazard. Multi Science Publishing Co. Ltd., Essesx

5. Kumar MD, Srinivasu VK, Bassi N, Trivedi K, Sharma MK (2009) Groundwater Management in Rajasthan: Identifying Local Management Actions. Institute for Resource Analysis and Policy, Hyderabad

6. Millennium Ecosystem Assessment (2005) Ecosystems and Human Well-being: Synthesis. Island Press, Washington DC

7. Narain P, Khan MA, Singh G (2005) Potential for Water Conservation and Harvesting against Drought in Rajasthan, India. Working Paper 104 (Drought Series: Paper 7). International Water Management Institute (IWMI), Colombo

8. Nicholls RJ, Lowe JA (2006). Climate stabilization and impacts of sea-level rise. In H J Schellnhuber, W Cramer, N Nakicenovic, T Wigley and G Yohe (eds). Avoiding Dangerous Climate Change Cambridge University Press, Cambridge 195–202.

9. Ozha DD, Sharma SK (2011). Mitigation of Water Quantity and Water Quality Challenges in Groundwater of Rajathan. Water Research and Development. 1(1):29.

10. Ramesh R, Yadava MG (2005) Climate and water resources of India. Curr Sci 89(5):818–824

11. Rathore MS (2005) Groundwater Exploration and Augementation Efforts in Rajasthan- A Review. Institute of Development Studies, Jaipur

12. Ribot Jesse C (1996) Climate Variability, Climate Change and Social Vulnerability in the Semi Arid Tropics. Cambridge University Press, Cambridge

13. Sen Roy S, Singh RB (2002) Climate Variability, Extreme Events and Agriculture Productivity in Mountain Regions. Oxford and IBH Publication, New Delhi

14. Sharma S, Bhattacharya S, Garg A (2006) Greenhouse gas emissions from India: a perspective. Curr Sci 90(3):326 333

15. Sharma HS, Sharma M (2002) Facets of Rajasthan. Kuldeep Publishers, Ajmer

16. Singh RB, Gurjar AK (2011) Climate change vulnerability of agriculture in arid region of Rajasthan, India. Annals of National Association of Geographers, India 31(1):31–38

17. Smith JB et al (1996) Considering Adaptation to Climate change in the Sustainable Development of Bangladesh. World Bank Report, World Bank, Washington DC

18. Thornthwaite CW (1948) An approach toward a rational classification of climate. Geogr Rev 38(1):55–94

19. Watson RT, Zinyoera MC, Moss RH (1996) Climate Change 1995: Impacts, Adaptations and Mitigation of Climate Change: Scientific-Technical Analysis. Contribution of Working Group II to the Second Assessment Report. Intergovernmental Panel on Climate Change. Cambridge University Press, Cambridge

# Index

**P**



**S**



**T**



**W**



**X**



**Z**